The ocean—atmosphere system

The ocean – atmosphere system

A. H. Perry and J. M. Walker

Longman
London and New York

Longman Group Limited London

Associated companies, branches and representatives
throughout the world

Published in the United States of America
by Longman Inc., New York

First published 1977

Library of Congress Cataloging in Publication Data

Perry, A H
    The ocean-atmosphere system.

    Bibliography: p.
    1. Ocean-atmosphere interaction. 2. Ocean
circulation. 3. Atmospheric circulation.
I. Walker, J. M., joint author. II. Title.
GC190.2.P47   551.4'7   76-44276
ISBN 0-582-48559-2
ISBN 0-582-48560-6 pbk.

Printed in Great Britain by
Lowe & Brydone Printers Limited, Thetford, Norfolk

Preface

This book is concerned with *interactions* between the atmosphere and the oceans and with the *interdependence* of atmospheric and oceanic circulations. Few books treat meteorology, climatology and physical oceanography in an integrated manner. Indeed, we know of none that does so at a level which satisfies the needs of those to whom this book is addressed, namely postgraduate students and second- and third-year undergraduates in the broad fields of geography, geophysics, environmental science, marine biology and nautical studies. The monographs of E. B. Kraus (1972), *Atmosphere—Ocean Interaction*, and H. U. Roll (1965), *Physics of the Marine Atmosphere*, are of an advanced nature, and to derive full benefit from them considerable competence in physics is required. Moreover, Kraus expects his reader to be familiar with fundamental principles of fluid mechanics. In contrast, *Atmosphere and Ocean*, by J. G. Harvey (Artemis Press, 1976), is intended as an introductory undergraduate text. We are aware of no other books which adopt a comprehensive integrated approach to atmospheric and oceanic science at university/polytechnic level. The gap in the available literature is thus apparent.

In describing the ocean—atmosphere system we have chosen to take a broad view which embraces phenomena and processes ranging in scale from the microscopic to the global. The intention throughout has been to provide a review of knowledge and understanding and to place theories and findings in their historical context. In selecting references we have taken care to give priority to review papers, to 'classics' and to reports and papers which discuss recent research. More than half the references cited are to works published during the last decade.

The book contains six chapters:

In Chapter 1 we introduce the principal features of the ocean—atmosphere system, we review the historical perspective, and we mention some of the means by which information about atmospheric and oceanic behaviour is obtained.

In Chapter 2 we seek an answer to the question, 'What causes ocean currents?' With the aid of case studies we find that currents in the uppermost layers of the ocean are largely wind-driven, whereas deep-water circulations are due primarily to thermohaline processes.

Physical processes occurring at the air—sea boundary and in the turbulent atmospheric and oceanic boundary layers are considered in Chapters 3 and 4. Thus, we elaborate upon mechanisms invoked in Chapter 2.

In Chapter 5 we investigate inter-relationships between anomalies of sea-surface temperature and behaviour-patterns of synoptic-scale weather systems, and we discuss the rôle of the oceans in climatic change.

Finally, in Chapter 6, we examine the aims and aspirations of international research projects currently in progress and draw attention to developments in numerical modelling of the ocean—atmosphere system. In so doing we assume that readers are familiar with the rudiments of mathematical modelling and that they realize the potentialities and limitations of the numerical approach. We feel that this assumption is reasonable, because most science-based undergraduates are required nowadays to undertake courses in computing, and these courses generally include modelling.

It gives us great pleasure to acknowledge those who have assisted in the preparation of this book. Professor R. G. Barry (University of Colorado) and Professor S. Gregory (University of Sheffield) read the entire work, Dr C. H. Cotter (University of Wales Institute of Science & Technology) read all but Chapter 5, and Mr L. Draper (Institute of Oceanographic Sciences) read the sections on waves and swell. We are most grateful for their helpful advice and constructive criticism. Any errors which may remain are unquestionably ours. We are indebted also to the library staffs of the University College of Swansea, the University of

v

Wales Institute of Science & Technology, the British Library and the Meteorological Office for their courtesy and efficiency, and we wish to record our thanks to Miss A. R. Bugden and her colleagues at the Institute of Oceanographic Sciences for providing such detailed replies to requests for oceanographic information. To conclude, we extend our gratitude to Mrs G. Whitehouse (University of Wales Institute of Science & Technology), who produced the typescript so meticulously.

Without the tolerance and forbearance of our respective wives, of course, the book could never have been written.

Allen Perry Malcolm Walker
Swansea *Cardiff*

July 1976

Abbreviations and acronyms

AIDJEX	Arctic Ice Dynamics Joint Experiment
AMTEX	Air-Mass Transformation Experiment
ASWEPS	Anti-Submarine Warfare Environmental Prediction System
ATEX	Atlantic Trade-Wind Experiment
BOMEX	Barbados Oceanographic and Meteorological Experiment
CAENEX	Complete Atmospheric Energetics Experiment
FGGE	First GARP Global Experiment
GARP	Global Atmospheric Research Programme
GATE	GARP Atlantic Tropical Experiment
GFDL	Geophysical Fluid Dynamics Laboratory
GISS	Goddard Institute for Space Studies
ICSU	International Council of Scientific Unions
IDOE	International Decade of Ocean Exploration
IGOSS	Integrated Global Ocean Station System
IGY	International Geophysical Year
IIOE	International Indian Ocean Expedition
IOC	Intergovernmental Oceanographic Commission
IQSY	International Quiet Sun Years
ITCZ	Intertropical Convergence Zone
JASIN	Joint Air–Sea Interaction Project
JOC	Joint Organizing Committee of GARP
JONSWAP	Joint North Sea Wave Project
LEPOR	Long Term and Expanded Programme of Oceanic Exploration and Research
MODE	Mid-Ocean Dynamics Experiment
MONEX	Monsoon Experiment
NCAR	National Center for Atmospheric Research
NORPAX	North Pacific Experiment
NSF	National Science Foundation
ONR	US Office of Naval Research
POLEX	Polar Experiment
SCOR	Scientific Committee on Oceanic Research
UNESCO	United Nations Educational, Scientific and Cultural Organization
WMO	World Meteorological Organization
WWW	World Weather Watch

Contents

Acknowledgements to figures

Fig. 1.3.   Reproduced from Shaw, 1926; Fig. 102, p. 259.

Fig. 1.4.   Reproduced from Groen, 1967; Fig. 95, p. 213.

Fig. 1.5.   From Bailey, 1953; p. 23 (figure not numbered).

Fig. 2.1.   Reproduced from Budyko, 1974; Fig. 2, p. 9.

Fig. 2.2.   After Gates, 1962; Fig. 6, p. 15.

Fig. 2.3.   After Sellers, 1965; reproduced from McIntosh & Thom, 1969; Fig. 3.4, p.31.

Fig. 2.4.   Based on the data of London, 1957; diagram reproduced from Neumann & Pierson, 1966; Fig. 9.13, p. 244.

Fig. 2.6.   Reproduced from McIntosh and Thom, 1969; Fig. 10.1, p. 186.

Fig. 2.7.   Reproduced from McIntosh and Thom, 1969; Fig. 10.2, p. 187.

Fig. 2.8.   Reproduced from Roberts, 1971; Fig. 149, p. 334.

Fig. 2.9.   After *The Library Atlas*, G. Philip & Son Ltd., 1967.

Fig. 2.10. Reproduced from Roberts, 1971; Fig. 150, p. 335.

Fig. 2.11. After *The Library Atlas*, G. Philip & Son Ltd., 1967.

Fig. 2.12. Data source: *Tables of Temperature, Relative Humidity and Precipitation for for the World*, Part III, HMSO, London, Met.O.617c, 1967.

Fig. 2.13. Data source: Neumann & Pierson, 1966; p. 156.

Fig. 2.14. Reproduced from Düing, 1970; Fig. 1, p. 4.

Fig. 2.15. Reproduced from Düing, 1970; Fig. 2, p. 5.

Fig. 2.16. Reproduced from Düing, 1970; Fig. 3, p. 9.

Fig. 2.17. Reproduced from Düing, 1970; Fig. 4, p. 9.

Fig. 2.18. After Düing, 1970; Fig. 6, p. 11.

Fig. 2.19a–e. After Düing, 1970; Figs. 9–13, pp. 22–6 respectively.

Fig. 2.20. After Groen, 1967; Fig. 96, p. 214.

Fig. 2.21. After Munk, 1950; Fig. 8, p. 89.

Fig. 2.22. Reproduced from Stommel, 1958; Fig. 2, p. 82.

Fig. 2.23. Reproduced from Groen, 1967; Fig. 112, p. 251.

Fig. 2.24. After Iselin, 1936; reproduced from von Arx, 1962; Fig. 11.2, p. 314.

Fig. 2.25. Reproduced from Groen, 1967; Fig. 114, p. 254.

Fig. 2.26. Reproduced from Groen, 1967; Fig. 115, p. 254.

Fig. 2.27. Reproduced from von Arx; Fig. 11.22, p. 342.

Fig. 2.28. After Groen, 1967; Fig. 39, p. 107.

Fig. 2.29. Traced from the facsimile charts broadcast from Bracknell on 4,782 kHz at 1400 GMT on the appropriate dates.

Fig. 2.30. After Walker & Penney, 1973; Fig. 3, p. 363.

Fig. 2.31. Based on data tabulated by Vowinckel & Orvig, 1970.

Fig. 2.32. Taken from Walker & Penney, 1973; Fig. 4, p. 365.

Fig. 2.33. Original, except that the outline of Antarctica and the spring ice-limit were traced from *The Library Atlas*, G. Philip & Son Limited, 1967.

Fig. 2.34. Data source: W. Schwerdtfeger (1970) 'The Climate of the Antarctic', in *World Survey of Climatology*, Vol. 14 (Ed. S. Orvig), Elsevier, 253–355.

Fig. 2.35. After Neumann & Pierson, 1966; Fig. 7.12, p. 149, who acknowledge US Naval Oceanographic Office, Washington DC.

Fig. 2.36. Taken from Kort, 1962; p. 86 (figure not numbered).

Fig. 2.37. Taken from von Arx, 1962; Fig. 5.5, p. 131.

Fig. 2.38. After Wyrtki, 1961; Fig. 3, p. 54.

Fig. 3.1.   Reproduced from Darbyshire & Draper, 1963; Fig. 1, p. 5.

ix

Fig. 3.2.   Reproduced from Darbyshire & Draper, 1963; Fig. 2, p. 6.
Fig. 3.4.   After Barber, 1969; Fig. 7.14, p. 136.
Fig. 3.5.   Taken from Mason, 1975; Fig. 8, p. 35.
Fig. 3.7.   Taken from Groen, 1967; Fig. 52, p. 131.
Fig. 3.8.   After Wu, 1959; Fig. 2, p. 446.
Fig. 3.9.   After Ekman, 1905.
Fig. 3.10. After Stewart, 1967; diagram b, p. 40.
Fig. 3.11. After Neumann & Pierson, 1966; Fig. 8.6, p. 195.
Fig. 3.12. After Neumann & Pierson, 1966; Fig. 8.8, p. 197.
Fig. 3.13. Reproduced from Neumann & Pierson, 1966; Fig. 8.10, p. 200.
Fig. 3.14. After Dietrich & Kalle, 1957; Table 3.
Fig. 3.15. Adaptation of Groen, 1967; Fig. 104, p. 234.
Fig. 3.16. Taken from Rossiter, 1954; Fig. 8, p. 397.
Fig. 3.17. After Hunt, 1972; Fig. 4, p. 121.
Fig. 4.1.   Based upon data contained in the Smithsonian Tables.
Fig. 4.2.   Based upon data contained in the Smithsonian Tables.
Fig. 4.3.   Taken from Budyko, 1974; Fig. 23, p. 150.
Fig. 4.4.   Taken from Budyko, 1974; Fig. 25, p. 153.
Fig. 4.5.   Taken from Budyko, 1974; Fig. 26, p. 155.
Fig. 4.6.   Data extracted from C. S. Ramage, F. R. Miller and C. Jefferies, *Meteorological Atlas of the International Indian Ocean Expedition*, Vol. 1, National Science Foundation, Washington DC, 1972; Charts, 55, 67, 127, and 139.
Fig. 4.7.   Taken from Budyko, 1974; Fig. 27, p. 158.
Fig. 4.8.   Taken from Budyko, 1974; Fig. 28, p. 159.
Fig. 4.9.   Taken from Budyko, 1974; Fig. 30, p. 164.
Fig. 4.10. Taken from Budyko, 1974; Fig. 31, p. 165.
Fig. 4.11. Taken from Roll, 1965; Fig. 81, p. 254, who based the diagram upon data published by Gordon, 1952.
Fig. 4.12. Taken from Budyko, 1974; Fig. 33, p. 168.
Fig. 4.13. Taken from Budyko, 1974; Fig. 34, p. 169.

Fig. 4.14. Taken from Budyko, 1974; Fig. 32, p. 166.
Fig. 4.15. Taken from Budyko, 1974; Fig. 35, p. 171.
Fig. 4.16.a–f. Taken from Budyko, 1956; Fig. 39, p. 129,   1974; Fig. 66, p. 203 1956; Fig. 41, p. 131   1974; Fig. 68, p. 205   1956; Fig. 43, p. 133 1974; Fig. 70, p. 206, respectively.
Fig. 4.17. Taken from Budyko, 1974; Fig. 37, p. 174.
Fig. 4.18. After Drozdov & Berlin, 1953; reproduced from Malkus, 1962; Fig. 14, p. 131.
Fig. 4.19. After Budyko, 1956, simplified version of Malkus, 1962, Fig. 13, p. 128.
Fig. 4.20. After Sellers, 1965; reproduced from Palmén & Newton, 1969; Fig. 2.4, p. 40.
Fig. 4.21. Graphs prepared from data tabulated by Newell *et al.*, 1970.
Fig. 4.22. Taken from Groen, 1967; Fig. 38, p. 105.
Fig. 4.24. Taken from Hubert, 1966; Fig. 4, p. 7.
Fig. 4.25. After Burt *et al.*, 1974; Fig. 7, p. 5629.
Fig. 4.26. Taken from Stevenson, 1968; Fig. 5, p. 160.
Fig. 4.27. Taken from Harrold & Browning, 1969; Fig. 7, p. 718.
Fig. 4.28. Taken from Palmén, 1948; Fig. 4, p. 31.
Fig. 4.29. Data source: *Meteorology for Mariners*, HMSO, Met.O.593, 1967, 304 pp.
Fig. 4.30. Taken from Walker, 1972b; Fig. 3, p. 171.
Fig. 4.31. Taken from Ramage, 1972; Fig. 1, p. 485.
Fig. 5.1.   After Perry, J. D., 1968; Fig. 1, p. 35.
Fig. 5.2.   Taken from Groen, 1967; Chart 2.
Fig. 5.3.   After Dietrich & Kalle, 1957; Table 3.
Fig. 5.4.   After Lumb, 1961; Plate 6.
Fig. 5.5.   After Smed, 1966; ICNAF special publ., vol. 6, p. 822.
Fig. 5.6.   After Vacnadze *et al.*, 1970; p. 36.
Fig. 5.7b. After Perry, A. H., 1968; pp. 252–3.
Fig. 5.8.   After Gagnon, 1964.
Fig. 5.9.   After Barry, R. G. and Perry, A. H., 1973, *Synoptic Climatology*; Figs. 5.51 & 5.52.
Fig. 5.10. After Wolff, P. M. & Laevastu, T. (1968) 'The effect of oceanic heat exchange on 500 mb large-scale pattern change', *Tech. Note No. 35*, Fleet Num. Wea. Fac., Monterey, 15 pp.; Fig. 1.
Fig. 5.11. After Namias, 1969; p. 189.
Fig. 5.12. After Bjerknes, 1969; p. 111.
Fig. 5.13. After Flohn, 1973; p. 135.

Fig. 5.14. After Lamb, 1972; Fig. 8.9, p. 334.

Fig. 5.15. After Lamb, 1966, *The Changing Climate*, Methuen; Fig. 15, p. 135.

Fig. 6.1. Information source: Kuettner & Parker, 1976.

Fig. 6.2. Taken from Manabe & Bryan, 1969; Fig. 1, p. 786.

Fig. 6.3. Taken from Manabe *et al.*, 1974; Fig. 4.8, p. 58.

Fig. 6.4. Taken from Shukla, 1975; Fig. 1, p. 504.

Fig. 6.5. After Worthington, 1977, reproduced from Bretherton, 1975; Fig. 4, p. 708.

Appendix diagram After Helland-Hansen, 1916; pp. 357–9.

Acknowledgements

We are grateful to the following for permission to reproduce copyright material:
Cambridge University Press for a figure giving 'The Bursting of Bubbles at the sea-surface' from *Clouds, Rain and Rainmaking* by B. J. Mason; The Royal Society for a chart by Halley giving 'Winds over Tropical Oceans' from *Phil. Trans.* 1688, Vol. 16; The Copyright Agency of the U.S.S.R. for figures by M. I. Budiko in *Climate and Life* 1974; Engineering magazine for two figures from an article entitled 'Forecasting wind-generated Sea Waves' by M. Darbyshire and L. Draper in *Engineering* 195, 1963; W. H. Freeman & Company for a map giving 'A General Outline of the Voyage of the Challenger' by Herbert S. Bailey Jr. in an article entitled 'The Voyage of the Challenger' in *Scientific American* May 1953. Copyright © 1953 by Scientific American Inc. and data giving 'The Antarctic Ocean' by V. G. Kort in *Oceanography* September 1972. All rights reserved; The Controller of Her Majesty's Stationery Office for two figures from *Handbook of Aviation Meteorology*, data from *Tables of Temperature Relative Humidity and Precipitation for the World Pt. III* 1967, redrawing of two charts showing 'Sea-surface Isotherms and Sea-Ice limits in the Greenland Area (a) 28.8.75 to 4.9.75 and (b) 21-15.2.76'. All reprinted by permission; The author for a figure from *Water Waves* by Norman F. Barber 1969; The authors for figures from *Essentials of Meteorology* by D. H. McIntosh and A. S. Thom; Institute of Oceanographic Sciences for data from *Phil. Trans.* Royal Society A246, 1954, paper by Dr. J. R. Rossiter; Pergamon Press Ltd. for a figure giving 'Multiple-current Interpretation of the data collected simultaneously at three widely separated ship tracks' by F. C. Fuglister in *Deep-Sea Res.*, 2, 1955; George Philip Printers Ltd. for an extract which has been redrawn giving 'Prevailing Surface Currents of the Oceans' based on a copyright map by George Philip & Son Ltd; Prentice-Hall Inc. and U.S. Naval Oceanographic Office for figures from *Principles of Physical Oceanography* by Gerhard Neumann and Willard J. Pierson Jr. © 1966. Reprinted by permission; Royal Meteorological Society for various figures from *Weather, Quarterly Journal of the Royal Meteorological Society* and *The Global Circulation of the Atmosphere* ed. by G. A. Corby.

Whilst every effort has been made to trace the owners of copyright, in a few cases this has proved impossible and we take this opportunity to offer our apologies to any authors whose rights may have been unwittingly infringed.

Chapter 1

The nature and characteristics of the ocean–atmosphere system

Although Greek philosophers in ancient times and Renaissance scientists in the sixteenth and seventeenth centuries were certainly acquainted conceptually with *systems*, 'sets of objects together with relationships between the objects and between their attributes' (Hall & Fagen, 1968), it is only in the years since the Second World War that the structures and properties of systems have been investigated intensively and their complexities realized. The outcome has been a flood of research papers and a great many books, some of which show that familiar fields of study can be transformed by the application of systems thinking (see, for example, Chorley & Kennedy, 1971). Moreover, out of the upsurge of interest in systems has evolved a new science, *cybernetics* (see Wiener, 1968), which is concerned with the mathematical structure of control and communication in electronic and in naturally-occurring systems. Nonetheless, the importance of explaining things in terms of systems has possibly been overstated, for an observation made by Ludwig von Bertalanffy in 1956 remains valid today: 'if someone were to analyse fashionable catchwords he would find "systems" high on the list'.

In geophysics and physical geography, the fields of study which most concern us in this book, the emphasis upon systems during the last three decades has not only come as a welcome contrast to the descriptive compilation approach formerly favoured but has also yielded considerable advances in the understanding of relationships between the atmosphere and the waters of the sea. Before we proceed to discuss these advances, however, we must introduce the concept of the ocean–atmosphere system, review the historical perspective and consider the means by which information about this system can be obtained.

Reasons for studying the ocean–atmosphere system

Like every other system, the ocean–atmosphere system consists essentially of elements and links between the elements and it regulates its own behaviour by the operation of internal feedback mechanisms.

Until recently, oceanographers and meteorologists focused their attentions on the elements of the ocean–atmosphere system and paid comparatively little heed to links and to feedback mechanisms. To quote Roll (1972):

For several decades oceanographers and meteorologists used to follow different avenues of research when investigating the structure and behaviour of their respective media. The former were mainly concerned with the processes in the oceanic depths, while the latter concentrated their interest on the upper layers of the atmosphere. The interface between ocean and atmosphere received but little attention. During the last two decades, however, the situation has changed remarkably, air–sea interaction studies now being 'in vogue' and rapidly increasing in number.

In several research institutes oceanographic and meteorological studies are combined, as, for example, at The Ocean–Atmosphere Research Institute in Cambridge, Massachusetts, and The Institute for Meteorology and Oceanography at Utrecht.

To those engaged in the pursuit of pure science, it is intellectual curiosity and the challenge of a difficult task which motivate endeavours to understand the ocean–atmosphere system. To applied scientists, though, and to those who allocate funds for research projects, there are other compelling reasons for investigating the system. For example, the World Ocean is becoming increasingly a theatre of military operations, so that various aspects of atmospheric and oceanic behaviour are now of concern to defence strategists (the factors affecting underwater sound propagation, to name but one aspect). Further, as Couper (1974) has emphasized, it is imperative that due consideration be given to the weather and sea conditions likely to be experienced during

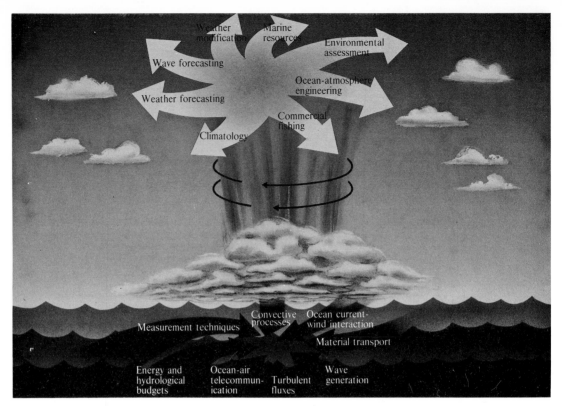

Fig. 1.1. Hidy's (1972) air–sea interaction swirl of utility.

exploration and exploitation activities connected with the development of offshore hydrocarbon resources, since safety to personnel, the risk of pollution and the possibility of destruction of expensive equipment are involved. Weather and wave forecasts are particularly important to the offshore industry. Improvements to the accuracy of these forecasts depend to a considerable degree upon greater understanding of the ocean–atmosphere system.

A comprehensive treatment of reasons for interest in the complex, but fascinating, ocean–atmosphere system would fill a book many times the size of this. For a list of reasons, reference should be made to Hidy (1972), who has drawn an analogy between the hurricane sucking up from the sea quantities of heat and water vapour and what he termed 'the air–sea interaction swirl of utility' (Fig. 1.1) sucking up from the 'sea of knowledge' and spinning out into many fields of application. Nevertheless, failure to mention pollution would be a serious omission.

Atmospheric and oceanic pollution is a matter of concern to everyone. More than half a million different substances are imposed upon the natural environment each year, and the sea is the ultimate sink for the bulk of them, including airborne gases and solids, since these are washed from the atmosphere by precipitation sooner or later. So effective are atmospheric and oceanic transports that pollution reaches all parts of the globe; even the penguins of Antarctica contain DDT in their fat (see Woodwell, 1967). However, the state of knowledge about transport and dispersal of pollution in the ocean–atmosphere system unfortunately leaves much still to be desired.

In recent years there has been much talk of harvesting the sea to help alleviate the problem of human starvation (see, for example, Ricker 1969), but there seems little point in directing much effort towards that end when the living resources which might be harvested are threatened by the toxic and carcinogenic substances which are present in the sea and the oil and non-degradable products of industry floating upon its surface. As Ehrlich and Ehrlich (1972) have written:

No one knows how long we can continue to pollute the seas with chlorinated hydrocarbon insecticides, polychlorinated biphenyls, oil, mercury, cadmium, and thousands of other pollutants, without bringing on a worldwide ecological disaster. Subtle changes may already have started a chain reaction in that direction, as shown by declines in many fisheries, especially those in areas of heavy pollution caused by dumping of wastes.

It is surely paradoxical that Man is striving to understand the ocean—atmosphere system so that he might maximize food production from the sea (see Budyko, 1974), yet at the same time he is interfering with ecosystems and thereby thwarting this ambition; however, Bascom (1974) takes an optimistic view of waste-disposal in the ocean.

The nature of the ocean—atmosphere system

A fundamental feature of ecosystems was versified by Jonathan Swift (1667—1745) in his *On Poetry:*

So, naturalists observe, a flea
    Hath smaller fleas that on him prey;
And these have smaller fleas to bite 'em,
    And so proceed *ad infinitum.*

This was paraphrased by Augustus de Morgan (1806—71) in *A Budget of Paradoxes:*

Great fleas have little fleas upon their backs to bite
    'em,
And little fleas have lesser fleas, and so *ad
    infinitum,*

which was adapted by the mathematician L. F. Richardson (1881—1953) with reference to atmospheric motions:

Big whirls have little whirls
    which feed on their velocity.

Little whirls have lesser whirls
    and so on to viscosity.

We have resisted the temptation to adapt further Swift's verse in respect of the interconnected skein of systems of different scales which constitutes the ocean—atmosphere system; rather, we return to scientific terminology by quoting Malkus (1962), who, outlining kinetic energy transfer between atmospheric and oceanic eddies of various sizes, has written of: 'the coexistence in each (fluid medium) of many interacting scales of motion from tiny eddy to planetary gyre, supplying and removing energy from one another, coupled in loops within loops of stable and unstable interaction'. The following extract from a paper by White and Barnett (1972) both introduces the nature of the principal interactions and demonstrates that the importance of feedback in the ocean—atmosphere system is appreciated nowadays:

Over the mid-latitude North Pacific, large temporal and spatial variations (on the order of years and thousands of kilometres, respectively) are found in the monthly and annual indices describing the general circulation of the atmosphere. However, the natural meteorological time scales are short compared to these index fluctuations. Therefore, the ocean may play an important role in the maintenance of these variations, acting to influence the atmosphere through the effect of the ocean circulation upon heat exchange between the two fluid media. This heat exchange can significantly affect the wind systems in the atmosphere. In turn, the circulation in the upper layers of the ocean is driven by the wind systems. The resulting mutual interaction can be thought of as a servomechanism acting to couple the two fluid media.

A servomechanism is a feedback control system.

Table 1.1   Scales in atmospheric and oceanic motions (after Roll, 1972)

Name	Characteristics	Dimensions	
		Time	Horizontal space
Microscale	Turbulent motion (molecular exchange at the sea surface)	Up to several minutes	Up to 100 metres
Convective scale	Pronounced vertical motion	Several minutes to 1 hour	100 metres to 10 kilometres
Mesoscale	Tendency toward organized motion	Several hours	10 to 100 kilometres
Synoptic scale	Cyclonic and anticyclonic vortices around vertical axes	Several days	100 to 1,000 kilometres
Large scale	Quasi-stationary circulations, planetary waves, anomalies of the 'Grosswetterlage', climatic fluctuations	Weeks to decades	1,000 kilometres to whole ocean

3

Scales of atmospheric and oceanic motion are summarized in Table 1.1. In general, the restlessness of the atmosphere, especially in the lower troposphere, is paralleled in the upper layers of the sea, but, on account of the density of sea-water being approximately one thousand times that of air near sea-level, there is a difference in space and time scales between similar atmospheric and oceanic phenomena. Oceanic fronts, for example, possess dimensions typically a few per cent of their atmospheric counterparts, and they maintain their identity for several weeks, compared with the several days characteristic of atmospheric fronts.

It is at the air–sea boundary and in the overlying atmospheric layer where the oceans and the atmosphere exchange energy, that such small-scale processes as, for example, wind-stress, heat transfer and evaporation take place. These processes are of fundamental importance in the generation and maintenance of all atmospheric and oceanic circulations. Stewart (1975) has put it thus:

All water vapour entering the atmosphere by evaporation from the surface must enter through the boundary layer. Sensible heat can and does enter and leave the atmosphere by a variety of paths and for this reason is more complex than is momentum or even water vapour (which is complex for other reasons because of the phase changes involved) but transfer of sensible heat through the boundary layer is frequently very important and in some cases dominant.

It should not be forgotten that the boundary layer plays another role as well, in particular with respect to climate. Through the boundary layer the ocean gains most of its momentum. There it also loses a substantial quantity of water, resulting in increased salinity at the surface. The resulting momentum input and density changes are largely responsible for the oceanographic circulation. The circulation of the ocean plays its role again on the atmosphere by moving heat from one portion of the globe to another, in particular and most importantly across latitude lines, assisted by the presence of meridional continental boundaries which do not have anything like a comparable effect on the atmosphere which they have on the ocean. The ocean is believed to have an effect on climate fully comparable to that of the atmosphere, although of course discussed in this way it is not really possible to separate their effects and one should talk of the joint ocean–atmosphere system.

However, as Stewart pointed out, 'study of the boundary layer is not only very important, it is also very hard! The difficulties are both conceptual and observational. The conceptual problem lies in the fact that with trivial exceptions the boundary layer is turbulent.' The air–sea boundary itself is a moving surface of such great complexity that it repels investigation. There is, consequently, much yet to be learned about processes in the upper ocean and in the atmospheric layer adjacent to the surface of the sea. We discuss the action of wind on the sea in Chapter 3 and ocean–atmosphere heat exchange in Chapter 4.

Across the air–sea boundary there are transfers not only of heat and momentum but also of solids and gases. Kraus (1972) has noted that 'the sea surface acts like a membrane which regulates the rate at which many substances circulate through the system that is our planet'. The study of these fluxes has become an increasing preoccupation of the climatologist; as Hare (1966) said, 'one of the outstanding changes in climatology has been the shift away from such parameters as temperature and relative humidity towards the measurement of fluxes. From taking snapshots and time-exposures of the atmosphere, we have discovered a zest to understand the mechanisms of energy and moisture exchange.' The area of the ocean surface is more than twice that of the land; energy exchange over that surface dominates world climate.

To sum up this section on the nature of the ocean–atmosphere system we may paraphrase the words of Benton *et al.* (1963): the atmosphere and the oceans constitute a single mechanical and thermodynamical system of two coupled fluids and they interact in a manner which is so complex that cause and effect cannot always be distinguished.

Adem's (1973) schematic representation of interactions in the ocean–atmosphere system is shown in Fig. 1.2. It was Aristotle (384–322 BC) who wrote: 'When we examine a thing in its details, it appears larger than when considered as a composite whole.' Whether or not this is true of the ocean–atmosphere system is debatable. It can be argued that the system is a good example of holism.

A sketch of the historical background

We now turn to the history of endeavours to understand the ocean–atmosphere system. We make no apology for devoting space to historical

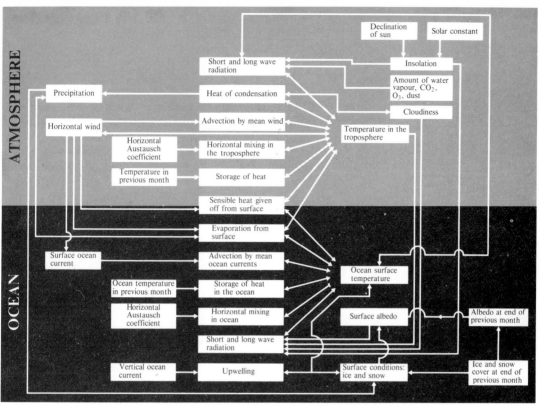

Fig. 1.2. Adem's (1973) schematic representation of interactions in the ocean–atmosphere system.

aspects of our subject. We consider it a matter of regret that the history of science is so neglected in curricula. As Forbes and Dijksterhuis (1963) have commented:

Not until people have learnt to see how our present knowledge and ability has grown from very modest beginnings, as a result of indefatigable collaboration between many minds and ceaseless collective development of what has once been achieved, will they be able to appreciate it fully and learn to use it with respectful gratitude.

According to Herdman (1923), 'oceanography is a subject of modern development, though of ancient origin'. This is true also of meteorology. Both fields of study have their origins in Greek science. Indeed, the word 'meteorology' is derived from the Greek τὰ μετέωρα, meaning 'the things in the air'; ὠκεανος (okeanos), to the Greeks, was the great stream or river supposed to encompass the Earth's disk, the Atlantic Ocean, the great outer sea beyond the Pillars of Hercules (as the mountains on the two sides of the Strait of Gibraltar were called).

The ancient Greeks were undoubtedly well aware of *interactions* between the atmosphere and the waters of the sea, for, in his celebrated treatise on atmospheric phenomena, *Meteorologica*, Aristotle postulated that water is evaporated from the surface of the sea by the action of the sun's rays, and he reiterated the hypothesis attributed to Anaxagoras of Clazomenae (*c.*500–428 BC), that water vapour rises into the air, where it is cooled and so condensed, and thereupon falls in the form of rain. Aristotle wrote: 'We always plainly see the water that has been carried up coming down again. Even if the same amount does not come back in a year or in a given country, yet in a certain period all that has been carried up is returned.' The hydrological cycle was thus recognized [1] .*

From the time of Aristotle to the Renaissance of science in the late Middle Ages very little progress was made towards an understanding of the ocean–atmosphere system. As Deacon (1971) has written in her admirable history of marine science:

---

* Numerals in [square] brackets indicate notes at end of the chapter.

Aristotle in particular dominated scientific thought until the seventeenth century and so wide was the range of his works and the influence of his opinion on all branches of learning that, until new work superseded his own, he was accredited with virtual omniscience. Any ideas about the sea were, true to this pattern, held to be of weight in the absence of more positive knowledge, if his name was attached to them, even if they were not in fact his own.

Geographical knowledge of atmospheric and oceanic behaviour increased down the centuries, however. The increase was slow during the mediaeval period, while only the Norsemen and the Arabs and various adventurous individuals (for example, the Irish monks who reached Iceland) were widening their horizons, but became much more rapid from the late fifteenth century onwards, in the 'Golden Age of Discovery', when sponsored voyages of exploration penetrated far beyond what was previously thought to be the limit of the known world. Geographical knowledge was essential if existing theories about the ocean–atmosphere system were to be tested and new ideas promulgated.

By the seventeenth century sufficient was known of the patterns of prevailing winds and ocean currents of the globe for the resemblance between the patterns to be apparent, but it seems that no one suggested explicitly that winds might be the cause of major ocean currents until 1699, when William Dampier (1652–1715) published the idea in his *Discourse of Winds, Breezes, Storms, Tides and Currents.* Hitherto, winds were believed to be responsible only for temporary movements of water. The pattern of ocean currents was commonly explained in terms of the hydrological cycle. In the words of Deacon (1971):

It was argued that the sun moving through the tropics evaporated water from the sea surface in these regions and lowered its overall level while rain fell nearer the poles causing a corresponding rise. Currents, which were by definition movements of water from a higher to a lower level, flowed from the poles towards the tropics and westwards along the equator in order to restore the equilibrium of the ocean.

The influence of Aristotle was clearly still very strong.

During the eighteenth century it was widely regarded as a self-evident truth that the great ocean currents depend upon prevailing winds, and in the early part of the nineteenth century Major

James Rennell (1742–1830) distinguished between *drift currents,* produced by the action of wind on the surface of the sea, and *stream currents,* formed when drift currents are deflected by obstacles. Redfield (1834), though, doubted the simple view that winds alone cause ocean currents, saying:

It is common to ascribe the currents of the ocean wholly to the action of the winds; but, as the waters of the ocean are subject to the same impulses as the superincumbent atmosphere, it is probable that the principal movements of both fluids have their origin in the same causes.

In his opinion, density contrasts between water masses provide some of the motive force behind major ocean currents.

It is surprising that the possibility of density differences initiating large-scale oceanic motions was neglected for so long, because the concept of a convective circulation in the Mediterranean Sea had been widely accepted since the late seventeenth century and a strong case for ocean circulations being maintained by density differences had been made in the late eighteenth century by the prominent scientists Richard Kirwan (1733–1812) and Benjamin Thompson, Count Rumford (1753–1814).

According to Kirwan (1787):

As the water in the high northern and southern latitudes is, by cold, rendered specifically heavier than that in the lower warm latitudes, hence there arises a perpetual current from the poles to the equator, which sometimes carries down large masses of ice, which cool the air, to a great extent.

Thompson (1798) noted that the density of pure water is greatest at about $40°$ F, whereas saline water is most dense at a somewhat lower temperature. He wrote:

As sea water continues to be condensed as it goes on to cool, even after it has passed the point at which fresh water freezes, the particles at the surface, instead of remaining there after the mass of the water had been cooled to about $40°$, and preventing the other warmer particles below from coming in their turns and giving off their Heat to the cold air (as we have seen always happens when fresh, or pure water is so cooled), these cooled particles of salt water descend as soon as they have parted with their Heat, and in moving downward force other warmer particles to move upwards; and in consequence of this continual succession of

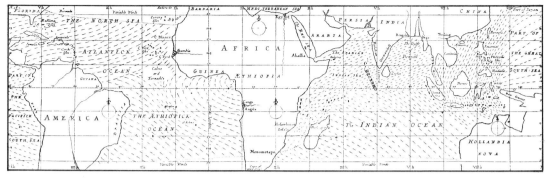

Fig. 1.3. Halley's chart of winds over tropical oceans.

warm particles, which come to the surface of the sea, a vast deal of Heat is communicated to the air.

The only source of heat for the sea is solar radiation, and the surface layers of the sea are chilled in high latitudes, mainly by the action of cold winds; therefore, Thompson concluded, water which

descends to the bottom of the sea, cannot be warmed where it descends, [and] as its specific gravity is greater than that of water at the same depth in warmer latitudes, it will immediately begin to spread on the bottom of the sea, and to flow towards the equator, and this must necessarily produce a current at the surface in an opposite direction; and there are the most indubitable proofs of the existence of both these currents.

In support of his hypothesis Thompson pointed out that cold water was known to be present at great depths in the ocean. Nevertheless, 70 years were to elapse before his manifestly plausible model of ocean circulation gained general acceptance.

An essentially similar model was proposed by Alexander von Humboldt (1769–1859). On his expedition to South America, his *Voyage aux régions équinoxiales du nouveau continent, fait en 1799–1804,* Humboldt (1814) found that 'dans les mers de Tropiques on trouve qu'à de grandes profondeurs, le thermomètre ne se soutient qu'à 7 ou 8 degrés centésimaux'. He explained: 'l'existence de ces couches froides dans les basses latitudes, prouve par conséquent un courant inférieur qui se porte des pôles vers l'équateur'. He continued: 'il prouve aussi que les substances salines qui altèrent la pesanteur spécifique de l'eau, sont distribuées dans l'Océan de manière à ne pas anéantir l'effect produit par les différences de température'[2].

We account for the circulation of the Mediterranean Sea in Chapter 2. At this point we merely grant to Henry Sheeres (d. 1710), Count Luigi Ferdinando Marsigli (1658–1730) and Edmond Halley (1656–1742) the credit for establishing that thermohaline processes drive the circulation, and note that Deacon (1971) has written at length about endeavours to explain currents in the Mediterranean Sea (particularly those in the Strait of Gibraltar). We now turn to trade-winds and associated ocean currents.

In the annals of science there have been few contributions to knowledge of the ocean–atmosphere system more notable than Halley's paper, *An Historical Account of the Trade Winds and Monsoons, observable in the Seas between and near the Tropicks; with an Attempt to assign the Phisical Cause of the Said Winds,* published in 1686 in *The Philosophical Transactions* of the Royal Society of London. Halley began his paper with a detailed and methodical description of surface winds over the Atlantic Ocean, the Indian Ocean, the western North Pacific Ocean and the eastern South Pacific Ocean (Fig. 1.3) and, in so doing, identified 'several problems, that merit well the consideration of the acutest naturalists'. He then dismissed the explanation of trade-winds, 'some have been inclined to propose', the notion that 'as the globe turns eastwards, the loose and fluid particles of the air, being so exceedingly light, are left behind; so that, in respect of the earth's surface, they move westwards, and become a constant easterly wind', a view held by such eminent philosphers as Galileo Galilei (1564–1642) and Johann Kepler (1571–1630); he wrote:

It remains therefore to substitute some other cause, capable of producing a like constant effect, not liable to the same objections, but agreeable to the

7

known properties of the elements of air and water, and the laws of the motion of fluid bodies. And such seems to be the action of the sun beams on the air and water, as he passes every day over the ocean, considered together with the nature of the soil, and situation of the adjoining continents. First then, according to the laws of statics, the air, which is less rarefied or expanded by heat, and consequently more ponderous, must have a motion towards those parts, where it is more rarefied and less ponderous, to bring it to an equilibrium; and secondly, the presence of the sun continually shifting to the westward, that part towards which the air tends, by reason of the rarefaction made by his greatest meridian heat, is with him carried westward, and consequently the tendency of the whole body of the lower air is that way. Thus a general easterly wind is formed, which being impressed on all the air of a vast ocean, the parts impel one another, and so keep moving till the next return of the sun, by which so much of the motion as was lost is again restored, and thus the easterly wind is made perpetual.

Halley thereupon earned the credit for being the first to recognize that the relative situation of land and sea is an important factor in the shaping of atmospheric circulation patterns, especially in the creation of monsoons. In attributing the westward component of trade winds to the effect of the sun shifting westward over the ocean, however, his intuition failed him. It remained for George Hadley (1685–1768) to explain (Hadley, 1735) that the westward component is due to the influence of the Earth's rotation on air currents flowing towards the equator (the Coriolis effect). Nevertheless, as Deacon (1971) has commented, 'his general explanation and the importance he attached to the relative situation of land and sea has not been challenged'. Certainly his convective concept was an improvement upon the notion of Martin Lister (*c.*1638–1712), who, in a paper read before the Royal Society in 1684, described the trade-winds as the 'constant breath' of the sargasso weed 'because the matter of that *Wind,* coming (as we suppose) from the breath of only one *Plant* it must needs make it constant and uniform: Whereas the great variety of *Plants* and *Trees* at land must needs furnish a confused matter of *Winds.'* Shaw (1926) has pointed out that this 'explanation' is evidently based upon Aristotle's theory of exhalation or emanation as the cause of winds. He commented: 'If Dr Lister had described the trade-winds as the surface-exhalation or emanation from

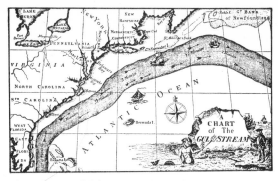

Fig. 1.4.   Franklin's chart of the Gulf Stream.

an anticyclone instead of from the plants that grow therein he would have made an effective contribution to the subject.'

Although Halley stated that his ideas about atmospheric circulation were based upon physical principles which apply equally to air and water, nowhere in his paper is there any reference to ocean currents. As we have seen, the possibility of there being convective circulations in the oceans was overlooked until the late eighteenth century. Indeed, for much of the eighteenth century there was rather little progress of any sort towards an understanding of the ocean–atmosphere system. Whereas marine science was pursued enthusiastically between 1660 and 1690, it was comparatively neglected during the next 60 years or so.

It was during the second half of the eighteenth century that scientific interest in the Gulf Stream was aroused. Much of the credit for stimulating this interest is due to Benjamin Franklin (1706–90), for:

(a) he published, in 1770, the first scientific chart of this current (see Ch. 2), depicting it as a river of warm water extending from the Florida Strait to far beyond the Newfoundland Banks (Fig. 1.4), a very much greater distance than hitherto realized (the existence of the Gulf Stream off Florida had been known for well over two centuries);

(b) he advanced the opinion that trade-winds drive the surface waters of the North Atlantic Ocean westward into the Gulf of Mexico, whence the water escapes through the narrow Florida Strait and thus forms the Gulf Stream (again see Ch. 2); and

(c) he was a strong advocate of the use of thermometers for ascertaining the direction of flow and the extent of the Gulf Stream and other currents.

During the late eighteenth and early nineteenth centuries knowledge of the Gulf Stream accumulated steadily. In particular, it came to be appreciated, from observations of sea—surface temperature, that the warm water which the current transports 'spreads itself out for thousands of square leagues over the cold waters around, and covers the ocean with that mantle of warmth which tends so much to mitigate, in Europe, the rigors of winter'. Maury (1844) continued:

Moving now more slowly, but dispensing its genial influences more freely, it finally meets the British islands. By these it is divided, one part going into the polar basin of Spitzbergen, the other entering the Bay of Biscay, . . . Such an immense volume of heated water cannot fail to carry with it, beyond the seas, a mild and moist atmosphere, and this it is which so much softens climate there. . . . Every west wind that blows crosses this stream on its way to Europe, and carries with it a portion of this heat to temper there the northern winds of winters.

The influence of the Gulf Stream upon the climate of north-west Europe was also stressed by Sabine (1846), who drew attention to his own suggestion, published in 1825, that the exceptionally mild weather experienced in Europe in the winter of 1821—2 was associated with a strengthening and 'extension of the Gulf Stream in that year to the coast of Europe, instead of its terminating as it usually does about the meridian of the Azores', and he ascribed the mildness of the winter of 1845—6 to similar behaviour of the current.

Sabine speculated further. He wrote: 'There can be little hesitation in attributing the unusual extension of the stream in particular years to its greater initial velocity, occasioned by a more than ordinary difference in the levels of the Gulf of Mexico and of the Atlantic.' Such a difference, he considered, results from an increase in strength of the trade-winds which drive tropical waters into the Gulf of Mexico. Bearing in mind that a time-interval of several months must elapse between tropical water receiving increased momentum from the trades and Gulf Stream water arriving off north-west Europe, he was led to conclude that there exists a correlation between the intensity of winter weather in Europe and the strength of trade-winds over the Atlantic Ocean during the preceding summer. He mused: 'how highly curious is the connexion thus traced between a more than ordinary strength of the winds within the tropics

in the summer, occasioning the derangement of the level of the Mexican and Caribbean Seas, and the high temperature of the sea between the British Channel and Madeira, in the following winter!'

It is probable that Sabine was the first to suggest that anomalies of atmospheric circulation might be related causally to oceanic anomalies elsewhere. The idea appears to have attracted little attention at the time, though, and the search for such relationships did not begin in earnest until the late nineteenth century. This search is still in progress, particularly for relationships between anomalies of sea—surface temperature and abnormal weather events remote in time and space. There have been disappointments, many enquiries producing inconclusive or negative results (see, for example, Bergsten 1936; Carruthers 1941), but in recent years climatologists have identified correlations apparently of practical value to long-range forecasters. We discuss anomalies in Chapter 5.

In the middle of the nineteenth century it was widely held that the waters of the Gulf Stream possess sufficient inertia to carry them to the shores of western Europe. This view was questioned by some, however, and by 1870 there was controversy over the cause of the warmth of the waters of the north-eastern North Atlantic Ocean. To a considerable extent this controversy was but one facet of a much greater controversy then at its height, that concerning the *raison d'être* of the general oceanic circulation.

Findlay (1869), for example, believed that the Gulf Stream *per se* cannot extend far beyond the longitudes of the Newfoundland Banks, where its waters mingle with the cold waters of the Labrador Current. He noted: 'The struggle between the Arctic and Tropical currents is here so strongly marked that the interlacing of the warm and cold waters, as shown by the thermometer, has been compared to the clasped fingers of the hands.' In his opinion, 'the Gulf Stream . . . can no longer be recognised beyond this cold-water gulf, which cuts off, as it were, its further progress, and which, it is manifest, it can neither bridge over nor pass under'. Yet, Findlay pointed out, there is evidence of the existence of the Gulf Stream well beyond the Newfoundland Banks, for 'cocoa-nuts and tropical produce are thrown upon the coasts of Iceland and Norway'. 'How then,' he added, 'can the phenomenon of our warm climate be accounted for?' The reason, he thought, was 'simple and obvious':

The great belt of anti-trade or passage winds which surround the globe northward of the Tropics, passing to the north-eastward, or from some point to the southward of west, pass over the entire area of the North Atlantic, and drift the whole surface of that ocean towards the shores of Northern Europe, and into the Arctic basin, infusing into high latitudes the temperature and moisture of much lower parallels; and which alone would be sufficient to account for all changes of climate by their variations, without any reference whatever to the Gulf Stream.

In the discussion which followed the presentation of Findlay's paper to a meeting of the Royal Geographical Society (on 8 February 1869), however, no less a personage than Professor T. H. Huxley showed himself to be unconvinced by Findlay's arguments. He considered that the latest maps showed without doubt that 'currents continuous with those of the Gulf Steam were traceable, with diminished velocity, to the northern points of the coast of Scotland'.

A different explanation for the warmth of the waters near the British Isles came from Carpenter (1870), who also 'expressed a doubt as to the extension of the Gulf Stream proper to the channel between the North of Scotland and the Faroe Islands'. He 'ventured to think it an open question whether the super-heating of the surface-water observed on a hot midsummer day beyond the northern border of the Bay of Biscay was not as probably due to the direct influence of the sun as to the extension of the Gulf Stream to that locality'. The north-eastward movement of 'the warm upper stratum of the North Atlantic,' he claimed, is 'part of a *general interchange* between Polar and Equatorial waters, which is quite independent of any such local accidents as those that produce the Gulf Stream proper, and which gives movement to a much larger and deeper body of water than the latter can affect'. He believed oceanic circulation to be essentially convective in nature.

Wyville Thomson (1871) was unable to accept the ideas of his friend William Carpenter and held to the view that 'the remarkable conditions of climate on the coasts of Northern Europe are due in a broad sense solely to the Gulf Stream'. He recognized that 'in a great body of water at different temperatures, under varying barometric pressures, and subject to the surface drift of variable winds, currents of all kinds variable and more or less permanent must be set up'; but he thought, nevertheless, 'the influence of the great current which we call the Gulf Stream, the reflux in fact of the great equatorial current, is so paramount as to reduce all other causes to utter insignificance'.

By the end of 1872 it was generally agreed that large-scale motions in the sea are caused partly by the impulse of winds and partly by contrasts of water density, and most scientists accepted that the relative magnitudes of the two driving forces could not be ascertained accurately without physically-based theoretical models of the ocean—atmosphere system and adequate data to quantify the models. As yet, theoretical models were but embryonic, and neither deep-sea soundings of temperature and salinity nor measurements of currents beneath the surface of the sea were available in sufficient numbers for hypotheses about oceanic circulation to be examined objectively[3]. After 1872 the controversy over the *raison d'être* of the general oceanic circulation degenerated into a personal quarrel between William Carpenter, who by now believed wind-driven and convection currents to be complementary, and James Croll, who insisted that *all* major ocean currents are wind-driven, even those in the Strait of Gibraltar. Deacon (1971) has fully documented this quarrel.

During 1872 marine scientists became preoccupied with preparations for an oceanographic expedition, a circumnavigation of the globe by the specially converted British corvette HMS *Challenger*. The object of her cruise was the scientific investigation of the deep ocean. In the doggerel of *Punch* (see Bailey, 1953):

Her task's to sound Ocean, smooth humours or
    rough in,
    To examine old Nep's deep-sea bed. . . .
In a word, all her secrets from Nature to wheedle,
    And the great freight of facts homeward bear.

Challenger sailed from Portsmouth on 21 December 1872, journeyed nearly 70,000 miles (Fig. 1.5), and returned to England on 24 May 1876. Tides were noted and currents measured; the physical and chemical properties of sea-water were studied; observations were made of meteorological and magnetic conditions; samples of oozes and other bottom deposits were dredged from the sea-floor; and specimens of flora and fauna were taken from the depths of the sea and from lands visited. Discoveries on the voyage were manifold, and so numerous were the results of the expedition that the official report on it comprised no fewer than fifty quarto volumes.

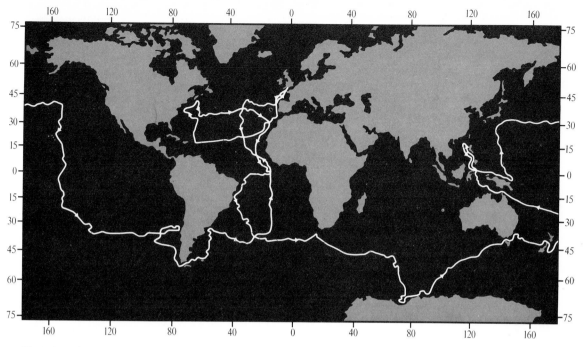

Fig. 1.5. The route of HMS *Challenger*.

From a biological stand-point the expedition was an enormous success, and chemists and geologists too were well satisfied. However, there was disappointment among those who hoped that physical observations made on the voyage would enable arguments over the cause of oceanic circulation to be resolved once and for all. The expedition lacked a physicist who could analyze the accumulated mass of data on temperature and specify gravity. Accordingly, so far as understanding of oceanic macro-circulations was concerned, the *Challenger* project represented a lost opportunity.

For a lucid account of the voyage of HMS *Challenger* and an assessment of the expedition's influence on the development of marine science in the next few decades, reference should be made to Deacon's book. We merely draw attention to the words of Bailey (1953): 'When (*Challenger*) had left England, the ocean deeps were an almost unfathomable mystery. When she returned, she had sounded the depths of every ocean except the Arctic and laid the foundation for the modern science of oceanography.'

Although the *Challenger* expedition was scientifically epoch-making, in that it effectively brought to a close the predominantly descriptive era of oceanography and ushered in the analytical era, there were in the middle of the nineteenth

century various other significant influences upon the course of marine science. Notable among these was the work of the distinguished American naval officer and hydrographer Matthew Fontaine Maury (1806–73), of whom Groen (1967) has written: 'He owes his high reputation not only to his own research at sea . . . but also, and pre-eminently, to his magnificent work in collecting, analysing and interpreting virtually everything relating to the sea, observed not only by scientists but notably by the seafarers.'

According to Roll (1965), 'the modern concept of considering ocean and atmosphere as a whole certainly has its source' in Maury's classic exposition of atmospheric and oceanic phenomena, *The Physical Geography of the Sea* (first published in 1855 [4]).

Undoubtedly Maury deserves due recognition for his services to practical navigation, for the charts of winds and ocean currents he published were of very considerable benefit to mariners. He also made several significant contributions to marine science, these being, in the words of Leighly (1968),

the instigation of uniform weather observations at sea according to a plan that was not outmoded until radio communication made it possible for ships to participate in regular synoptic observations;

11

the construction of the first maps of the surface temperature and of the bathymetry of the Atlantic Ocean; and the collection of samples of sediment from the bottom of the deep sea.

Nonetheless, we must dispel any notion which may still exist that Maury was a scientist. As Leighly (1968) has explained:

Maury was not content with putting useful information into the hands of practical men. The observations he collected demanded physical explanation, which he was bold enough to attempt. To his efforts at their interpretation he brought a lively imagination and unlimited self-confidence, but only the most superficial knowledge of physical science. This combination of qualities led him into grandiose but often fantastic generalizations concerning the circulation of the atmosphere and the oceans, which were justly rejected by his scientific contemporaries. . . . His theoretical interpretations may be most charitably left to oblivion.

Leighly does, however, give Maury 'credit for one indirect service to oceanographic science': William Ferrel (1817–91) was so provoked by reading Maury's fantastic interpretations of atmospheric circulation that he was moved to apply his knowledge of theoretical mechanics to the problem of fluid motions on a rotating body. The result was a series of outstanding papers on atmospheric and oceanic circulations, which Leighly considered 'the best in print for years after 1860, perhaps until the appearance of Henrik Mohn's essay of 1885' (Mohn, 1885). We refer again to Ferrel's work in Chapter 2.

At this point, having sketched the course of marine science from its origins to the beginning of the era when studies of the ocean—atmosphere system became truly scientific and endeavours to satisfy data requirements became systematic, we conclude our historical survey. Significant developments in marine science since the late nineteenth century receive mention elsewhere in the book. We now discuss the acquisition of data.

Marine observations

By studying the sailing directions which accompanied Maury's charts of winds and ocean currents (Maury, 1851) mariners were able to shorten their passages by many days. The navigational value of the charts was thus

demonstrated. Seafarers quickly realized that they would benefit not only from improvements to the charts but also from increased knowledge of all aspects of the ocean—atmosphere system. Accordingly, they willingly undertook to observe and note weather and sea conditions, and to furnish Maury with copies of their logs in exchange for sets of wind and ocean charts which were supplied to them gratis by the US Hydrographic Office which Maury superintended. Maury considered it desirable that observations be made and recorded according to a uniform plan and so, in 1853, on his initiative, a conference attended by delegates of several maritime nations was held (in Brussels) to devise a code of observational practice (for a brief account of this conference, see Maury 1855).

The delegates agreed that observations should be made every 2 hours and entered in a register under the following headings (see Parkhurst, 1955):

latitude and longitude (by observation and dead reckoning);
currents (direction and rate);
magnetic variations observed;
winds (direction, and force on the Beaufort scale);
barometer (height);
hours of fog, rain, snow and hail;
state of sea;
water (temperature at surface, specific gravity, temperature at depth);
state of weather;
remarks on tempests, tornadoes, whirlwinds, typhoons, hurricanes, waterspouts, temperature of rain, description of hailstones, dew, fog, dust, height of waves, tide-rips, colour of ocean, soundings, ice, shooting stars, Aurora Borealis, halos, rainbows, meteors, birds, insects, fish, seaweed, driftwood and tidal observations.

The interval between observations was soon altered from 2 to 4 hours, but the scheme has otherwise not been modified significantly to this day.

Seafarers and scientists alike derived benefit from the accumulating mass of marine observations. Seafarers took into account climatic normals when planning routes, and they also learned to recognize the warning signs of approaching storms. Scientists were provided with data for testing hypotheses. As Roll (1965) has noted: 'In maritime meteorology, as in any branch of natural science, suitable measurements are required as a basis for empirical investigations and as a touchstone for theoretical studies'.

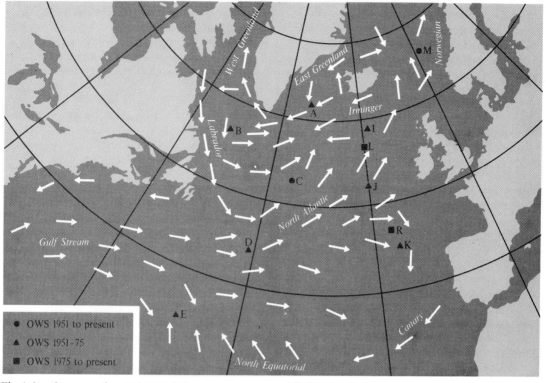

Fig. 1.6. Ocean weather stations and the principal currents of the North Atlantic Ocean.

At present, observations of the ocean—atmosphere system are made from about 4,000 merchant vessels of all nations. Nevertheless, observational coverage of the oceans is still far from adequate (see WMO, 1969). The majority of the vessels follow well-defined commercial sea routes, so that extensive tracts of ocean are rarely visited, and, for various reasons, reports are received from fewer than about 1,000 vessels on any given day. Roll has commented that 'a sufficiently dense and quasiuniform coverage is attained only in certain parts of the North Atlantic Ocean and its adjacent seas'. Nearly half of all marine observations are made on the North Atlantic Ocean and only about 10 per cent are made in the predominantly oceanic Southern Hemisphere. Except on the most frequented parts of the North Atlantic Ocean the density of synoptic observing stations on the oceans is even lower than in tropical deserts and on polar ice-caps. Moreover, weather routeing of ships across the North Atlantic and North Pacific Oceans has been practised increasingly in recent years, as a result of which observations in conditions of heavy weather have tended to become disproportionately few (see Quayle, 1974).

The *quality* of data derived from merchant vessels can be questioned, too, because (*a*) mariners generally receive rather little training in observational procedures and in the use of instruments, and (*b*) ships disturb mechanically the ocean and the atmosphere near them and also form sources of convective and radiative heat, factors which detract from the value of observed results (see Stevenson, 1964).

Since the Second World War a fixed network of ocean weather stations has been maintained on the North Atlantic and North Pacific Oceans, the work of the stations consisting of weather observations, aerological and hydrospheric soundings, and scientific programmes as circumstances permit. Roll has noted:

Complete series of meteorological observations, which were not possible before, can be obtained for fixed locations in the oceans over long periods. The weather ships' data constitute material of eminent value for nearly all studies in maritime meteorology. Their importance will even be increased in future years when the available period, which now comprises 14–17 years, will be longer.

Alas, the pressures of economics have lately been

such that there are now only four stations on the North Atlantic Ocean (see Meteorological Office, 1976), whereas there were nine when Roll wrote those words (see Fig. 1.6).

In recent years increasing interest has been shown in the development of floating automatic weather stations, for it is hoped that these might eventually be relied upon to supply data from oceanic regions at present poorly observed (see WMO/ICSU, 1974). In the United Kingdom and the United States especially, development of operational buoys is well advanced (see, for example, Day, 1976). However, the marine environment is hostile, and the problems of designing equipment which is both resistant to corrosion and sufficiently robust to withstand gales and heavy seas have yet to be overcome.

To obtain *detailed* information about atmospheric and oceanic phenomena special research enterprises are necessary. The *Challenger* expedition was the first major venture of this kind. In the century which has elapsed since that expedition the scientific cruise has become a familiar part of marine research programmes. Indeed, several countries (for example: the United Kingdom, the United States and the Soviet Union) today possess fleets of well-equipped research vessels, and research enterprises at sea have become progressively more ambitious. Projects nowadays involve international collaboration and the simultaneous operation of research vessels and aircraft. We discuss these projects in Chapter 6.

For many years reconnaissance aircraft have been used for surveying the marine environment, especially for locating and tracking tropical cyclonic storms. During the last 20 years or so, however, a new and effective means of observing and measuring atmospheric and oceanic phenomena has been developed, namely the artificial earth satellite. Weather forecasters and research workers already make great use of satellite observations of cloud distributions, and it is expected that satellite measurements of various elements of the ocean–atmosphere system will soon be available as a matter of routine (see, for example: Rao *et al.*,

1972; Badgley *et al.*, 1969; Houghton & Taylor, 1973; Gloersen & Salomonson, 1975). Furthermore, marine data can be retrieved from buoys by means of satellite interrogation. Undoubtedly the advent of satellites will prove one of the more significant technological advances in the progress of marine science.

Deacon (1971) has written:

Accurate measurement is a prerequisite of oceanography. The variations of temperature and salinity in the surface water are generally too small and too gradual to be detected by the unaided senses. Waves and tides are too ephemeral to be studied without a record while the movement of a current out of sight of land is imperceptible to the ship carried with it. When it is a question of investigating the characteristics of the great mass of the sea, below the surface, man's dependence on instrumentation becomes absolute.

A similar statement could be made about atmospheric phenomena. We should, though, bear in mind the words of Ludlam (1966a): 'observations alone reveal little without ideas to shape them'. Prior to about 1850 there were ideas but few observations. Nowadays there are abundant observations, there are computers to process data, there are ideas in plenty, and there are theoretical models. Yet, insight into the ocean–atmosphere system is still far from complete. This book is devoted to a study of progress towards the achievement of this insight.

Notes

1. For a thorough summary of Aristotle's *Meteorologica*, see Shaw (1926).
2. For historical details of the quest for an understanding of deep-sea circulations, reference should be made to Wüst's (1968) review paper.
3. For a discussion of the scientific results of expeditions made between the years 1749 and 1868, see Prestwich (1875).
4. This book has recently been reissued, with a critique by John Leighly (Maury, 1963).

Chapter 2

Oceanic macro-circulations

The energy which maintains atmospheric and oceanic motions is derived from the sun in the form of electromagnetic radiation. Accordingly, it is appropriate to begin this chapter with an examination of the global radiation budget, whence stems an understanding, albeit elementary, of the necessity for these motions. Only broad principles are considered at this stage, however; geographical and temporal radiation patterns receive detailed attention in Chapter 4.

Solar radiation

The distribution of energy with temperature (T) and wavelength (λ) for a black-body (i.e. perfect) radiator can be obtained from Planck's Law:

$$E_\lambda = \frac{C_1 \lambda^{-5}}{\exp\left(\dfrac{C_2}{\lambda T}\right) - 1}$$

where E_λ is the energy emitted in unit time from unit area within unit range of wavelength centred on λ, and C_1 and C_2 are constants. The wavelength of maximum energy emittance (λ_{max}) is given by Wien's Law:

$$\lambda_{max} T = 2,897 \ \mu\text{m K}$$

Observations reveal that the extraterrestrial solar spectrum is characteristic of a black-body radiating at a temperature of about 6,000 K, λ_{max} being 0·47 μm.

During passage through the atmosphere the solar beam is reduced (see Fig. 2.1). Radiation of wavelength shorter than 0·29 μm is absorbed totally by oxygen and by ozone in the high atmosphere and that of wavelength longer than 0·7 μm is absorbed selectively by oxygen in the high atmosphere and by water vapour in the troposphere. Although the atmosphere is transparent to radiation of the remaining wavelengths of the spectrum the solar beam is

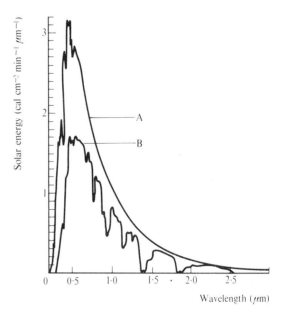

Fig. 2.1. Energy distribution in the solar spectrum. Curve A represents the extra-terrestrial spectrum and curve B the spectrum observed at ground level.

further reduced by scattering from air molecules, water vapour and dust.

Across a surface normal to the sun's beam the radiation flux at the fringe of the atmosphere (the *Solar Constant*) is 139·6 mW cm^{-2}. However, as the solar beam is intercepted by a circle of area πr^2, where r is the radius of the Earth, but distributed daily over the surface of a sphere of area $4\pi r^2$, the *mean* flux normal to the Earth's surface is 34·9 mW cm^{-2}. The Earth's albedo (reflection coefficient) is about 0·36, so that only some 22·5 mW cm^{-2} remain available for heating. The conversion from radiant to heat energy which accompanies the absorption of radiation of wavelengths shorter than 0·29 μm and longer than 0·7 μm consumes about 1 mW cm^{-2} in the high atmosphere and 4 mW cm^{-2} in the troposphere, leaving 16·5 mW cm^{-2}, comprising both direct and

15

diffuse (scattered) radiation, to be absorbed at the Earth's surface.

Terrestrial radiation

In the long term there is no evidence that the Earth as a whole is either warming or cooling progressively so a balance must exist between incoming solar and outgoing terrestrial radiation. To balance the incoming radiation amounting to $22 \cdot 5$ mW cm^{-2} the Earth should (applying the Stefan–Boltzmann Law: $E = \sigma T^4$[1], where E is the intensity of radiation per unit area of a black-body at temperature T and σ, Stefan's constant, is $5 \cdot 67 \times 10^{-9}$ mW cm^{-2} K^{-4}) radiate to space as a black-body of temperature about 250 K. However, the mean temperature of the Earth's surface is in fact about 288 K. Not only is there the obvious difference of nearly 40 K between these temperatures but also substitution of T = 288 K in the expression $E = \sigma T^4$ gives the value E = 39 mW cm^{-2}, which indicates that the Earth's surface emits more than twice as much energy as it receives (16·5 mW cm^{-2})!

For T = 288 K the majority of the radiation emitted from the Earth's surface lies in the wavelength range 4 to 80 μm, and λ_{max} is near 10 μm. Although radiation of some wavelengths, notably those between 8 and 12 μm, is able to pass directly to space unless intercepted by cloud, most of the outgoing terrestrial radiation is absorbed by the atmosphere, the gases principally responsible being water vapour and carbon dioxide (see Fig. 2.2).

According to Kirchoff's Law, a body which strongly absorbs radiation of a certain wavelength also strongly emits radiation of this wavelength. Thus, the outgoing terrestrial radiation is successively absorbed and re-emitted as it propagates upwards, but, because temperature decreases with height, slightly less radiation is re-emitted than is absorbed at a given level. Consequently there is a net flux of radiation towards space. The bulk of the water vapour and carbon dioxide is in the lower troposphere, and, taking a global average, the Earth effectively radiates to space from a level near 500 mb (approximately 5·5 km above sea-level).

Radiation balance

When downward-transmitted infra-red radiation

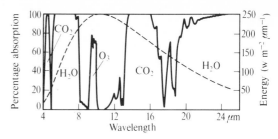

Fig. 2.2. Absorption curves for water vapour, carbon dioxide and ozone in relation to the radiation curve for T = 288 K.

from the atmosphere and from clouds is taken into account it is found that the Earth's surface as a whole possesses a large positive radiation balance; that is, the surface radiates less energy than it absorbs. Indeed, only in polar regions in winter is there a negative balance. On the other hand, the troposphere everywhere radiates more energy than it absorbs (see Fig. 2.3). To redress this imbalance water vapour and sensible heat are transferred by turbulence and convection from the surface into the troposphere, the water vapour releasing latent heat when condensation occurs.

For the surface and the atmosphere combined the radiation balance is positive equatorwards of latitude 38° and negative elsewhere (Houghton, 1954; Malkus, 1962). Since low latitudes are not steadily warming, nor high latitudes steadily cooling, horizontal advection of energy must take place (see Fig. 2.4). This is performed by wind systems in the atmosphere and by currents in the oceans, but the relative importance of these two media to the global heat balance, although long debated (see, for example, Neumann & Pierson, 1966), has still not been resolved indisputably.

The general circulation of the atmosphere

The manner in which redistribution of heat is accomplished by the atmosphere and the oceans is complicated. Consider first the essentially convective troposphere, within which Ludlam (1966b) has distinguished four fundamental kinds of convection:

Small-scale, nearly vertical convection, provides the lower troposphere with energy, partly as sensible heat and partly as the latent heat of evaporated water, and often manifests itself in the form of cumulus clouds. *Cumulonimbus* convection is much deeper than small-scale convection and is in low latitudes the principal

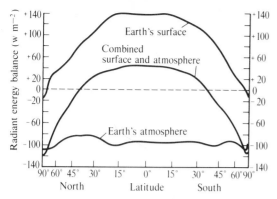

Fig. 2.3. Variation of radiation balance with latitude.

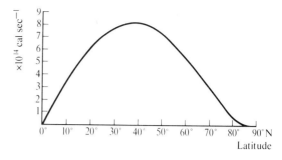

Fig. 2.4. The poleward transport of heat required by radiation estimates to maintain present climatic conditions on the globe.

means by which heat is distributed through the whole troposphere. In contrast, *large-scale* slantwise convection conveys heat both vertically and from low to high latitudes in the depressions and anticyclones of middle latitudes; in this kind of convection air motions are nearly horizontal and, unlike those in small-scale convection, are significantly subject to the effect of the rotation of the Earth (the Coriolis effect; see Coriolis, 1835). *Intermediate-scale* convection, on the other hand, possesses, to quote Ludlam, 'horizontal dimensions up to the continental and consists of baroclinic[2] circulations associated with irregular distributions of small-scale convection, imposed by the topographical features of the under-lying surface'. He cited as familiar examples sea-breeze and mountain-breeze circulations.

The general circulation of the atmosphere, which may be regarded as the synthesis of these convective motions, has been described lucidly both in articles (see, especially, Eady, 1964) and in numerous texts on elementary meteorology and has also formed the subject of an impressive monograph by Lorenz (1967). Accordingly, an outline will suffice here.

Briefly, the general circulation consists of three distinct, but collaborative, cells (Figs. 2.5 and 2.6):

1. In low latitudes a solenoidal Hadley cell prevails (Hadley, 1735). Near the equator heat is transferred from lower to higher levels of the troposphere by cumulonimbus convection (see, particularly, Riehl & Malkus, 1958). Polewards to approximately latitude $30°$ there is widespread subsidence of air throughout all but the lowest kilometre or two of the troposphere. At the surface, trade-winds — north-easterly in the Northern Hemisphere and south-easterly in the Southern Hemisphere — complete the cell. The trades of the two

hemispheres converge in a zone near the equator, the so-called 'Intertropical Convergence Zone' or 'ITCZ', where the deep cumulonimbus activity is concentrated.

2. In middle latitudes a Ferrel cell operates (Ferrel, 1889). Poleward of the quasi-permanent subtropical anticyclones (typically centred near latitude $30°$) weather is dominated, winter and summer, by transitory depressions and anticyclones. Although in a given locality considerable day-to-day variations of wind direction may occur, in association with the ever-changing nature of these weather systems, westerlies prevail at all levels in the troposphere.

3. In polar regions radiative cooling causes air in contact with the Earth's surface to contract. This in turn causes subsidence of air in the lower troposphere and an anticyclonic tendency. Surface air currents, flowing away from the poles, are deflected by Coriolis effect to become easterlies. These features of the polar cell are, however, inclined to be intermittent and indistinct, and averaging is necessary to reveal the easterlies and the high pressure. Even regions in very high latitudes are visited by depressions, but the intensity of attendant fronts generally declines poleward of approximately latitude $60°$.

The idealized pattern of pressure and winds over the surface of the globe is shown in Fig. 2.7.

What causes ocean currents?

When, in the eighteenth century, it was recognized that there is a striking similarity between the arrangement of major currents in the upper ocean and the pattern of prevailing winds in the atmosphere above (Figs. 2.8–2.11) the theory that wind provides the motive force for major ocean

17

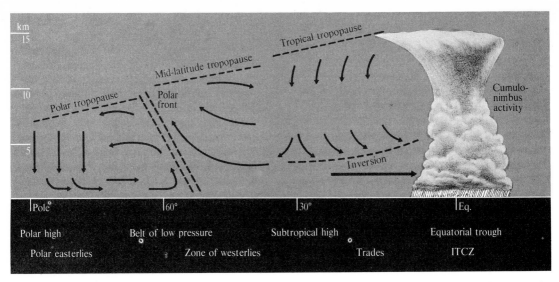

Fig. 2.5. Schematic representation of the atmospheric general circulation, in vertical section from polar regions to the equator.

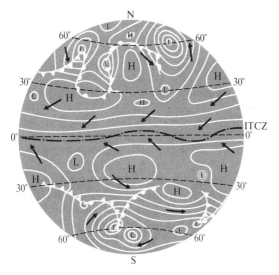

Fig. 2.6. Typical distribution of pressure systems, winds and fronts near an equinox.

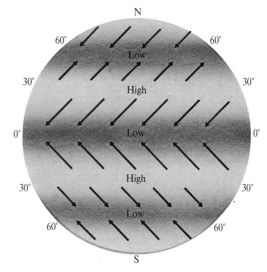

Fig. 2.7. Idealized pattern of pressure and winds over the surface of the globe.

currents became widely accepted. Not everyone was convinced, though, despite the dynamical support provided by Ferrel (1856, 1859–60, 1861). Indeed, by about 1870, a fierce controversy developed, as to whether ocean currents are due to the impulse of winds or to differences of specific gravity between the waters of the equatorial and polar regions (for an outline of the controversy see Chapter 1). As yet, none but the first hesitating steps had been taken towards a mathematical treatment which might settle the matter conclusively.

Otto Pettersson (1904) thought it 'obvious that the primary effect (of the prevailing winds) is the generation of surface currents of great intensity, and that the intensity of motion must decrease with depth'. Nevertheless, he considered that 'a great – and probably the greatest – part of the ocean current system must be due to the thermodynamic cycle of latent heat, consisting in the formation of ice in polar regions and the melting of ice in sea-water at lower latitudes', and he tried to justify his hypothesis with laboratory simulations involving the melting of blocks of ice in sea-water (Pettersson, 1904, 1907). The experimental work of Sandström (1908) and the theoretical treatment by V. Bjerknes (1916) indicated, however, that thermohaline circulations in the ocean are weak, unless equatorial waters can be heated to great depths. Soundings of these waters show that this does not happen, the reason being that heating by absorption of solar radiation is confined to a rather shallow surface layer. Nonetheless, predominantly thermohaline circulations do exist in the oceans, at least on a limited scale; witness, for example, water movements in the Mediterranean Sea.

Global wind and pressure patterns migrate latitudinally towards the pole of the summer hemisphere. Thus, the Mediterranean basin, which lies between latitudes 30° and 40°N, comes under the influence of extra-tropical disturbances in winter and subtropical subsidence in summer. Much of the annual precipitation falls in the winter part of the year (Fig. 2.12); in summer, since skies are largely cloudfree, precipitation is scanty and rates of evaporation are high.

There is a consequent net loss of fresh water from the Mediterranean Sea because precipitation and inflow from rivers are insufficient to counterbalance evaporation. Accordingly, Atlantic Ocean surface water flows through the Strait of Gibraltar to make good the deficiency. Moreover, since evaporation causes the salinity, and hence the density, of the surface water in the Mediterranean Sea to increase, convection develops. Dense surface water sinks, being replaced by less dense water, and eventually a bottom current flows out of the Mediterranean basin over the sill at the Strait of Gibraltar. A greater mass of water flows from the Atlantic Ocean into the Mediterranean Sea than flows back (Fig. 2.13), the difference being the balance of precipitation, evaporation and run-off from land.

Pettersson (1904) also suggested a relationship between outbursts of ice from Antarctica and fluctuations of monsoon rainfall in India. The suggestion apparently attracted little attention, even at a time when it was fashionable to attempt to establish relationships between monsoons and atmospheric and oceanic conditions elsewhere (see Normand, 1953; Walker, 1972a), and is worthy of mention here merely for the inherent irony of the reference to monsoons in a paper devoted to emphasizing the importance of thermohaline processes, when it is in the monsoon-influenced waters of the Arabian Sea that the significance of wind-stress in creating ocean currents is so

dramatically evident.

Winds and currents of the Indian Ocean

The semi-annual alternation of surface winds over the seas near southern Asia (Fig. 2.14) has long been known to mariners. The ancient Greeks knew from the campaigns of Alexander the Great; early Arab sailors made use of the winds on trading voyages along the coasts of Arabia and East Africa; and by the end of the first century AD Graeco-Roman seamen had discovered how to use the winds to cross the Arabian Sea. The associated reversal of surface currents in the North Indian Ocean (Fig. 2.15) appears to have been noticed in the eighth or the ninth century AD, according to Warren (1966) and Aleem (1967). Yet, despite this awareness of the reversals of wind and currents, there was, until the scientific awakening in the seventeenth century, little attempt to study monsoon events over the sea for any purpose other than to expedite trading.

By the end of the nineteenth century the general characteristics of the winds and surface currents of the Indian Ocean, particularly of the Arabian Sea and the Bay of Bengal, were known in some detail, but there was still a paucity of information on subsurface responses to the seasonal wind-reversal. Between 1857 and 1952 at least sixteen oceanographic expeditions visited the Indian Ocean. Even so, by 1957, to quote Currie (1966), who listed them, 'no large-scale systematic surveys had been made there and it was unquestionably one of the less well-known parts of the world ocean'. The International Indian Ocean Expedition (IIOE), which extended over the period 1959 to 1965, sought to rectify this situation, and of its objects Currie wrote:

Outstanding among the problems posed, the complete reversal of the wind system of the northern Indian Ocean with the change of the monsoon, seemed to present a unique opportunity of studying the reaction of the sea surface circulation to the wind. Such questions as how quickly the change in slope of the density layers responded and how quickly the subsequent biological events developed, were complementary to more general studies of these processes in the ocean. The probable upwelling regions which had been identified as developing seasonally during the south-westerly monsoon presented an opportunity of studying the process of upwelling as it develops. The meteorologist could see opportunities of

19

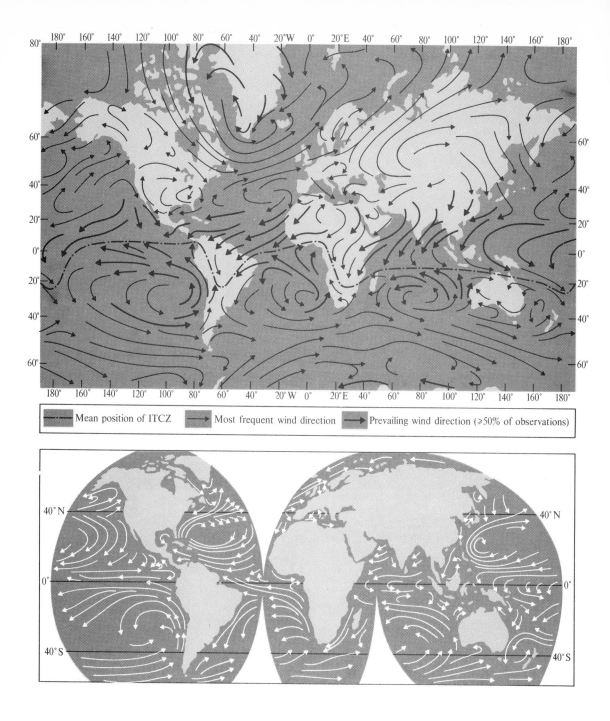

Mean position of ITCZ   Most frequent wind direction   Prevailing wind direction (≥50% of observations)

Fig. 2.8. Prevailing surface winds and the mean position of the ITCZ: January.

Fig. 2.9. Prevailing surface currents of the oceans: January.

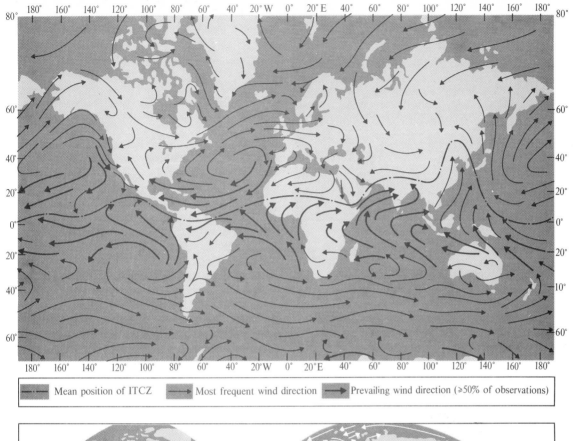

| Mean position of ITCZ | Most frequent wind direction | Prevailing wind direction (≥50% of observations) |

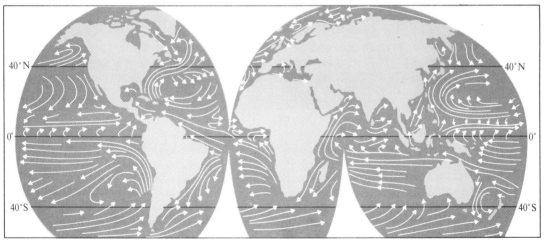

Fig. 2.10. Prevailing surface winds and the mean position of the ITCZ: July.

Fig. 2.11. Prevailing surface currents of the oceans: July.

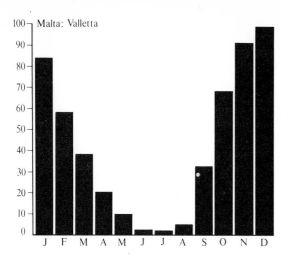

Fig. 2.12. Monthly—mean precipitation totals (mm) at Malta (Valletta).

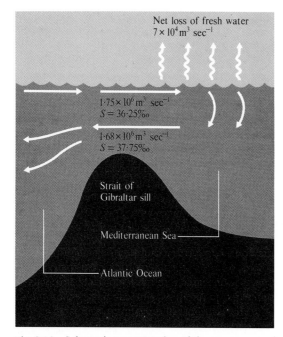

Fig. 2.13. Schematic representation of the average annual water balance of the Mediterranean Sea; S denotes salinity and the arrows indicate water movements.

learning a great deal about the monsoon circulation, the importance of which to the understanding of the climate of southern Asia hardly needed emphasis.

Many facets of air—sea interaction are exemplified in monsoon circulations of the Indian Ocean. Since these circulations have been explored systematically and in detail so recently, it is instructive to discuss monsoons at length.

Air-flows over the Arabian Sea

The credit for the first generally acceptable physical explanation of the Asian monsoon is due to Halley (1686). He realised that the rôle of convection is important (see Ch. 1) but his concept of the monsoon as a giant sea-breeze, although still widely accepted, is an over-simplification, for it fails to recognize that the monsoon is a *system*, in which convective processes of various scales *interact.* The nature of the interactions has been discussed by Walker (1972b, 1975), and the meteorology of monsoons has been thoroughly reviewed by Ramage (1971).

An important element of the convective circulations, in the context of the ocean—atmosphere system, is the south-westerly flow in the lower troposphere over the Arabian Sea. Since the time of Halley there has been widespread belief that this flow originates in the trade-wind belt of the winter hemisphere and, after being deflected by the Earth's rotation, enters the rain systems over western India. Some, notably Flohn (1953), have disputed this, pointing out that the turning of trade-winds into westerlies takes place to the south of the equator over much of the Indian Ocean. Only west of the meridian of $55°$E is there significant mass-transfer of air across the equator. Pisharoty (1965), during the IIOE, showed that the amount of water vapour transported into the Indian peninsula across its west coast on a day in July is two or three times the quantity transported across the equator, between the meridians embracing the Arabian Sea. He found also that the concept of cross-equatorial flow could not be supported from vorticity considerations and he concluded that most of the air up to a height of 6 km flowing eastwards across the Indian peninsula originates in the Northern Hemisphere; this latter conclusion has been supported by subsequent studies, although Saha (1970) believes less than 50 per cent originates in the Northern Hemisphere (see also Saha & Bavadekar, 1973). Moisture is added to the south-westerly air-flow from the underlying ocean by means of atmospheric turbulence (see Colon, 1964). Evidently there is, over the Arabian Sea, not only momentum transfer from the air to the sea but also latent heat transfer from the sea to the air. These aspects of air—sea interaction are discussed in detail in Chapters 3 and 4.

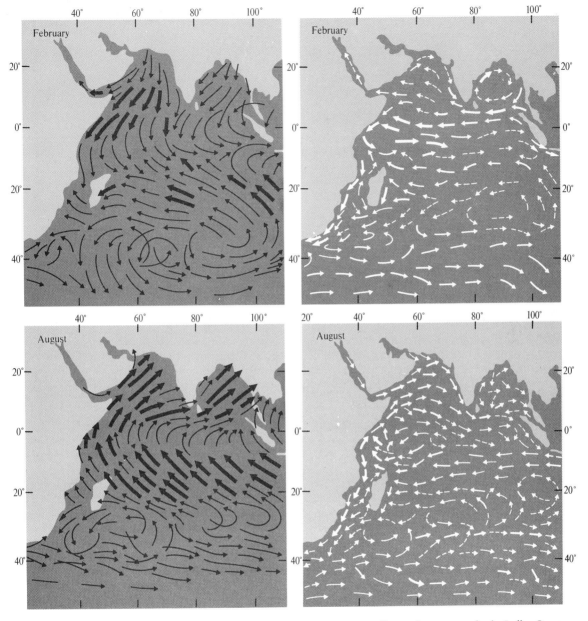

Fig. 2.14. Prevailing surface winds over the Indian Ocean in February and August.

Fig. 2.15. Prevailing surface currents in the Indian Ocean in February and August.

The extent of the monsoon influence on the Indian Ocean

It is an elementary feature of the scientific method that observations of an atmospheric or oceanic system must be obtained if a mathematical formulation, the ultimate expression of the behaviour of such a system, is to be verified. A great deal of effort in the IIOE was, of necessity, purely hydrographic. Simultaneously theoreticians developed mathematical models.

Düing (1970) has discussed at length the horizontal and vertical extents of the monsoon régime, basing his analyses on observations collected during the IIOE.

In determining the horizontal extent he scrutinized separately the winds and the surface currents.

23

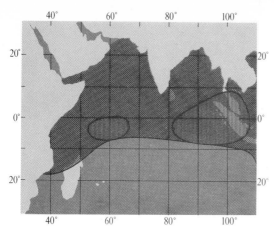

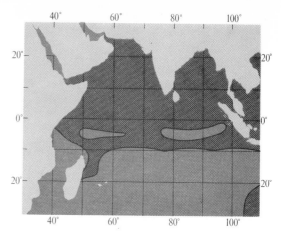

Fig. 2.16. Wind direction as a criterion of monsoonal influence. The diagonally hatched area indicates a change of more than 90° in mean wind direction between January and July.

Fig. 2.17. Surface current direction as a criterion of monsoonal influence. The hatched area indicates a change of more than 90° in mean surface current direction between January and July.

An area was considered to be monsoonal if the prevailing wind-direction changes by more than 90° between winter and summer, but he disregarded regions of light winds or calms (the doldrums), where directions can fluctuate widely over short intervals of time and distance. Defined in this way the monsoon is seen to be a phenomenon essentially of the Northern Hemisphere (Fig. 2.16); of wind patterns elsewhere over the Indian Ocean Düing wrote: 'Only in a few localities off the east coast of Africa and around northern Australia is the overwhelming, modifying influence of the ocean on the seasons overcome by the ocean–continent interaction that produces the monsoons'.

Likewise, he defined as monsoonal those areas where the direction of average surface currents changes by more than 90° between winter and summer (Fig. 2.17), but he pointed out that this definition is not entirely satisfactory, for two reasons: first, surface currents are only partly wind-driven. Western parts of the North Indian Ocean experience an excessively dry climate and eastern parts a rather humid climate, so there are considerable differences of surface salinity and temperature between these areas. Hence, the surface currents, although essentially wind-driven, are modified by thermohaline influences; and second, knowledge of surface currents is based largely on estimates of the drift of ships, the difficulty with this being that a ship drifts under the combined effect of wind and current.

Investigation of the vertical extent of monsoonal influences presented rather more

difficulty. As Düing put it: 'Owing to the lack of direct current measurements, such an investigation must be based on considerations of the dynamic heights[3] as computed from temperature and salinity observations.' The procedure adopted, again quoting Düing's words, was as follows:

First, the annual variation of the dynamic heights, ΔH, was determined by computing the mean values of the dynamic heights for each 10-degree square north of 20°S. The computations were carried out every month for the levels 0, 100, 150, 200, 300, 400, and 500 dbars. When enough data were available, the means were computed for 5-degree squares also. Although averaging over even smaller areas would have been desirable in the western boundary region of the ocean, this was not feasible, owing to the insufficient number of observations. The annual variation, ΔH, was then considered to be represented by the difference between the maximum mean value and the minimum mean value.'

Second, the depth of penetration of ΔH was determined. For this purpose, ΔH = 7·5 dyne cm was selected as a lower limit for a noticeable annual variation.

The vertical extent of the monsoonal influence within the Indian Ocean, so derived, is shown in Fig. 2.18. However, caution must be exercised when attempting to interpret Fig. 2.18 because a number of factors are involved. Not only have thermohaline and wind-driven circulations to be considered, but also dynamic processes and

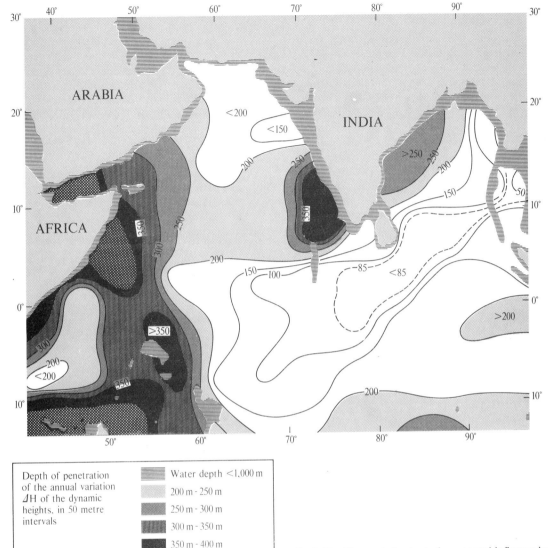

Fig. 2.18. The vertical extent of monsoonal influence in the waters of the Indian Ocean.

Depth of penetration of the annual variation *⊿H* of the dynamic heights, in 50 metre intervals

- Water depth <1,000 m
- 200 m - 250 m
- 250 m - 300 m
- 300 m - 350 m
- 350 m - 400 m
- >400 m

advection of water masses, as well as interactions between all these factors, must be taken into account. Figure 2.18, therefore, is at best only an approximate definition of the vertical extent of the monsoon regime.

The topography of the Indian Ocean surface

Having delineated the monsoonal influence Düing turned his attention to temporal variations of currents and sea-surface topography.

In the initial stages of the IIOE it was intended that monthly-mean charts of oceanic parameters

be drawn. However, on account of temporal and spatial heterogeneity of data, this proved impracticable, so it was decided instead to study characteristic periods. Two distinct atmospheric circulation patterns were recognized: (*a*) *winter* (December, January and February), when a Hadley-type circulation operates over the North Indian Ocean, with north-easterly trade-winds at the surface, and (*b*) *summer* (May to September inclusive), when the troposphere is reorganized and south-westerlies replace the north-easterlies. In the latter period, early summer (May–June) and late summer (July–September) were further distinguished. Spring and autumn, the transition

25

periods, were considered to be March—April and October—November, respectively.

Dynamic topography maps of the ocean surface were prepared, using 1,000 dbar as a reference level. These show the topography to be complex, rather more so in spring, early summer and autumn than in winter or late summer (Fig. 2.19), and quite unlike the sea-surface topographies of the Atlantic and Pacific Oceans, where, in contrast, large oceanwide gyres are conspicuous. Indeed, as Düing points out, 'the conditions prevailing in the northern Indian Ocean are found nowhere else'.

Internal waves, half-daily and daily tides, and inertial factors were rejected as significant influences on the sea-surface topography of the Indian Ocean. However, the upward slope of the sea-surface, from the western to the eastern part of the ocean between about 8°S and 20°N (see Fig. 2.19), seems to be related more to salinity gradients caused by climatic conditions in the overlying atmosphere than to wind stress. Over the western part of the Arabian Sea the climate is arid and, correspondingly, evaporation rates are high. Over the eastern part and over the Bay of Bengal conditions are more humid, and there is copious precipitation. Furthermore, surface salinities in the Bay are affected by plentiful drainage from rivers. No east—west gradients are obvious in the dynamic topographies for a depth of 100 m so this climatic effect appears to be restricted to a shallow surface layer of the ocean.

Finally in his discussion of IIOE data, Düing related dynamic topographies to the most prominent current systems of the tropical Indian Ocean, namely the North and South Equatorial Currents, the Countercurrent, the South-West Monsoon Current and the Somali Current.

The Somali Current

During the south-west monsoon there is in the upper 200 m or so of the ocean a strong north-eastward flow parallel to the coast of eastern Africa, close inshore, and, according to Swallow and Bruce (1966), about 250 km wide. This by some is still known as the *East Africa Coast Current* but oceanographers nowadays seem to prefer to name it the *Somali Current*. Warren *et al.* (1966), for example, find the latter name 'more descriptive [than the former] because the current does in fact flow mainly along the coast of Somalia, and the term "East Africa" is being used less for this part of the world as time goes on'. In autumn,

in response to the return of north-easterly trade-winds over the northern Indian Ocean the Somali Current reverses and thenceforth flows south-westward.

The Somali Current is considerably stronger during the summer monsoon than during northern winter. Indeed, between July and September surface flows of more than 3.5 m sec^{-1} have been observed off the Horn of Africa, south of Socotra, and in those months surface speeds exceeding 2.0 m sec^{-1} seem to be usual from near the equator to beyond Socotra (Swallow & Bruce, 1966). Moreover, Swallow and Bruce have estimated that the volume flow of the Somali Current during summer is of the order of 50×10^5 m^3 sec^{-1}. Accordingly, in the words of Lighthill (1969a), this current is 'too strong to be a merely local response of the ocean to the local winds' and 'it tends to be interpreted rather as part of the ocean's dynamic response to the pattern of wind stress over a large part of it'.

The expansion of knowledge resulting from the IIOE served to encourage theorists to develop mathematical models of Indian Ocean circulations. Prior to the IIOE there were insufficient data, especially on conditions beneath the surface, for comparing theory with reality. It is not appropriate in this chapter to dwell on such models and attendant assumptions, approximations, limitations and errors, but reference ought, however, to be made to some at this point.

The second part of Düing's (1970) monograph is devoted to a theoretical interpretation of the peculiarities of monsoonal circulations. To this end Düing achieved some measure of success and, in particular, he was able to show that westward intensification of wind-driven currents in the Indian Ocean occurs only in the presence of frictional dissipation.

Using a sophisticated three-dimensional numerical model with irregular lateral and bottom boundaries Cox (1970) also successfully simulated the large-scale features of the currents and water masses of the Indian Ocean, and he, too, reproduced the Somali Current and its seasonal variation, managing in so doing to predict fairly accurately its structure and seasonal phase in relation to the driving winds.

Lighthill's (1969a, b) theoretical model of the dynamic response of the Indian Ocean to the onset of the south-west monsoon is particularly impressive and replicates observed characteristics of the Somali Current in some detail. He applied the linearized theory of unsteady wind-driven currents in a horizontally-stratified ocean to an equatorial ocean with a western boundary and concluded (Lighthill, 1969b) 'that "wave packets" of current pattern reaching such a boundary deposit the "flux" they carry (velocity normal to the boundary integrated along it) in a boundary current which rather rapidly takes a rather concentrated form'. His theory indicates that the Somali Current forms within a month of the onset of the south-westerly monsoon winds, a time-scale which agrees well with observations.

Using the same type of theory Veronis and Stommel (1956) estimated that the time-scale for the response of the baroclinic part of a mid-latitude ocean with a western boundary to changes of wind-stress pattern is of the order of decades. This is in marked contrast to the month or so estimated by Lighthill in respect of the baroclinic Somali Current. However, Lighthill (1969b) believes that proximity to the equator, through latitude-dependent terms in his equations, might explain why the western Indian Ocean responds to wind-stress changes so much more rapidly than a mid-latitude ocean.

The origin of the Gulf Stream

Near the western boundary of the North Atlantic Ocean a strong surface flow extends, year in year out, from the subtropics well into middle latitudes (Fig. 2.20). This, in loose terminology, is the *Gulf Stream*. More precisely, following the nomenclature of Iselin (1936), the entire set of western currents, comprising the Florida Current (between the Florida Strait and Cape Hatteras), the Gulf Stream proper (between Cape Hatteras and the tail of the Grand Banks) and the North Atlantic Current (the continuation of the Gulf Stream eastward from the Grand Banks), is known as the *Gulf Stream System* (Fig. 2.20).

Once it was believed that floods of the Mississippi River produce the flow, but, according to Maury (1858), 'Captain Livingston overturned

Fig. 2.19a—e. The dynamic topography of the surface of the Indian Ocean (a) in spring, (b) in early summer, (c) in late summer, (d) in autumn and (e) in winter. Dots indicate observation points.

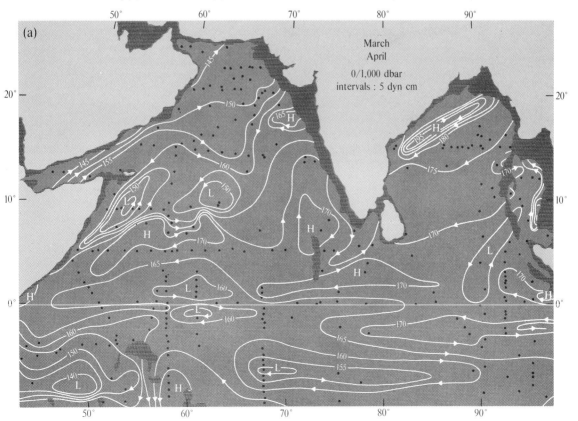

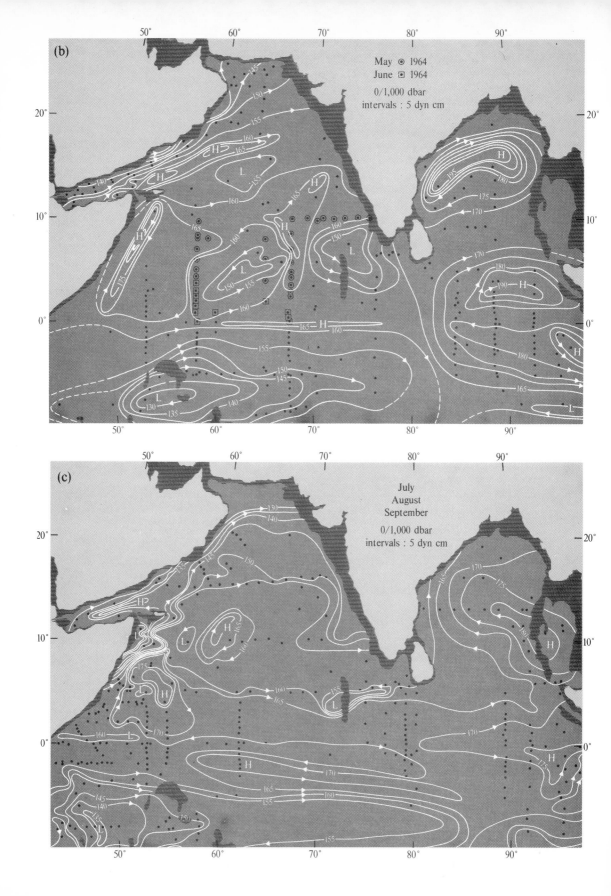

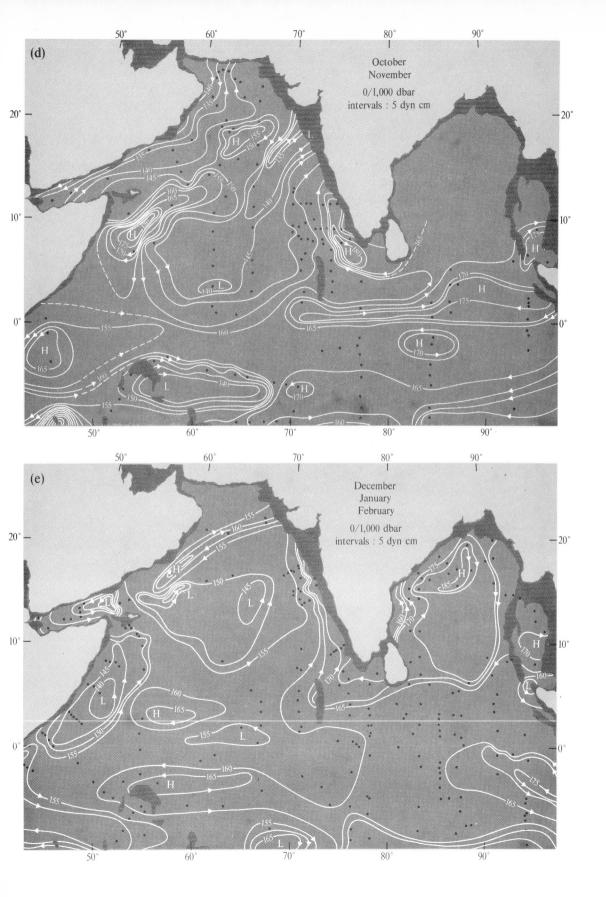

(d)

October
November

0/1,000 dbar
intervals : 5 dyn cm

(e)

December
January
February

0/1,000 dbar
intervals : 5 dyn cm

29

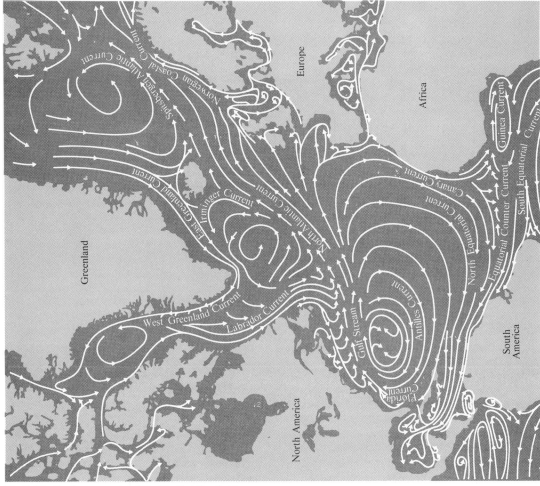

Fig. 2.20. Surface currents of the North Atlantic Ocean.

this hypothesis by showing that the volume of water which the Mississippi River empties into the Gulf of Mexico is not equal to the three thousandth part of that which escapes from it through the Gulf Stream'. More plausible was the opinion which held, again in the words of Maury, 'that the Gulf Stream is the escaping of the waters that have been forced into the Caribbean Sea by the trade-winds, and that it is the pressure of those winds upon the water which drives up into that sea a head, as it were, for this stream'.

The existence of this process is acknowledged today. North-east trade-winds drive the waters of the Atlantic Ocean slowly westward in the form of the North Equatorial Current. Near 60°W the current divides. A large proportion, together with a portion of the South Equatorial Current, flows into the Caribbean Sea and the remainder passes to the north of the West Indies as the Antilles Current

(Fig. 2.20). The water which flows into the Caribbean Sea is driven on into the Gulf of Mexico, whence it returns to the Atlantic Ocean through the Florida Strait. At its narrowest the Strait is only 80 km wide. Thus, concentration of streamlines occurs and a swift current is produced hydraulically, speeds in excess of 2 m sec^{-1} being not unusual. The average water transport through the Strait is 26×10^6 m^3 sec^{-1}. In the Florida Current reinforcement by the Antilles Current increases the transport to some 38×10^5 m^3 sec^{-1}, and further reinforcement by water of high salt content from the Sargasso Sea swells the transport to a peak of about 74×10^6 m^3 sec^{-1} in the Gulf Stream near 38°N.

Most of the water which enters the Gulf Stream System is water previously driven westward by the trade-winds. However, the processes just outlined do not account for the maintenance of intensely-crowded streamlines and mean speeds well in excess of $1 \cdot 0$ m sec^{-1} throughout the whole length of the western boundary of the North Atlantic Ocean from Florida to the Grand Banks. Credit for explaining this phenomenon physically is due chiefly to Stommel (1948) and Munk (1950).

Stommel's contribution was the demonstration from an analytical study of the wind-driven circulation in a homogeneous rectangular ocean under the influence of a simple distribution of surface wind-stress, linearized bottom friction, horizontal pressure gradients (caused by variable surface height), and Coriolis force, that the intense crowding is caused by variation of the Coriolis parameter[4] with latitude. He recognized the artificial nature of his model, saying that he thought this work 'suggestive, certainly not conclusive'. In particular, his assumption of a homogeneous ocean is at variance with observational evidence, for the majority of the transport in the principal currents takes place in the uppermost few hundred metres of the ocean; it also obliged him to adopt a somewhat arbitrary representation of bottom friction.

Munk was stimulated by Stommel's discovery and by the work of Rossby (1936) and Sverdrup (1947) to resume Ekman's (1923, 1932) attempts to account theoretically for the main features of ocean circulations. He assumed mean wind-stresses on the ocean surface and frictional dissipation forces within the ocean more typical of a real ocean than assumed in earlier models; and, with reference to a rectangular ocean basin, he successfully reproduced the arrangement of ocean circulations in the northern and southern hemispheres. Munk concluded that 'the circulations in the upper layers of the oceans are the result chiefly of the stresses exerted by the winds' and that his investigation had served to emphasize the fundamental importance of wind *torque*, rather than of the wind-stress *vector*, in determining the transport of ocean currents in meridionally-bound oceans. His model, however, yielded an estimate of the water transport in the Gulf Stream considerably smaller than that observed. This discrepancy he attributed to an underestimate of the stress at wind-speeds of less than about 7 m sec^{-1}. The problem of formulating wind-stress on an ocean surface is discussed in Chapter 3.

For a succinct exposition of the dynamical reasons for westward-intensification a review paper by Longuet-Higgins (1965) may be consulted. The skeleton of the argument is as follows:

(a) Absolute vorticity[5] (the sum of *planetary* vorticity arising from the rotation of the Earth and vorticity *relative* to the Earth's surface) can be equated to frictional forces plus applied wind-stresses. Viscous friction and relative vorticity can be neglected in comparison with planetary vorticity and wind-stresses. Hence, this relation can be derived:

$$M_y \frac{\partial f}{\partial y} = \text{curl}_z \, \tau$$

where M_y is the meridional component of the total horizontal mass transport and y the meridional distance, both positive northwards; f is the Coriolis parameter; and curl$_z \, \tau$ is the applied wind-stress torque about a vertical axis, cyclonic positive[6]. The relation is due to Sverdrup (1947).

(b) Within the subtropics, where there are trade-winds, and in middle latitudes, where there are westerlies, curl$_z \, \tau$ is negative. Accordingly, in the Northern Hemisphere, water must move southwards (M_y negative) because $\frac{\partial f}{\partial y}$ is positive.

(c) Mass must be conserved, so that there must be a return flow somewhere in the ocean. Vorticity must also be conserved; to achieve this, cyclonic vorticity must be generated somewhere. These requirements are satisfied by a current flowing rapidly northwards along the western boundary of the ocean, with lateral friction at this boundary (see Fig. 2.21).

Thus-formed are the western-boundary flows found in the North Pacific Ocean (the Kuroshio Current), the southern Indian Ocean (the Agulhas Current) and the South Atlantic Ocean (the Brazil Current)[7].

Water transport in the Gulf Stream is several times that in the Brazil Current. This, according to Munk, is due largely to the wind-stress curl being greater over the North Atlantic Ocean than over the South Atlantic Ocean. In Munk's words:

Differences in wind-stress curl can be ascribed to the northward displacement of the climatic equator relative to the geographical equator, which reduces the distance between the trades and the westerlies in the northern hemisphere to about 60 per cent of that in the southern hemisphere. The displacement of the trades in turn seems to be related to the unequal distribution of land and sea in the two hemispheres. One is tempted, then, to trace the preponderance of the great Gulf Stream and Kuroshio systems over their southern

31

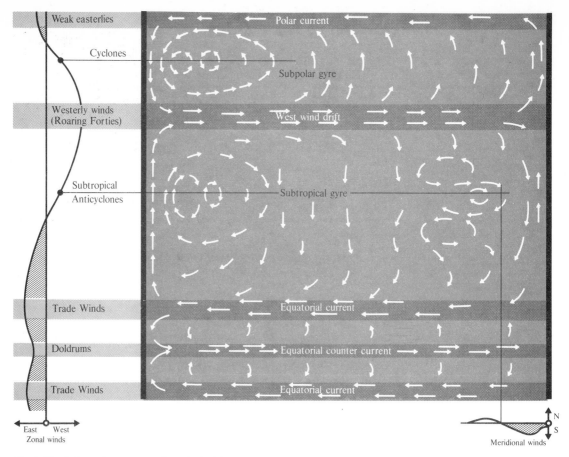

Fig. 2.21. Schematic representation of wind-driven currents in the upper layers of a rectangular ocean.

counterparts to the excess of water-covered area in the southern hemisphere.

Munk and Carrier (1950) improved upon Munk's model by considering triangular and semi-circular ocean basins and showed thereby that the inclination of the western boundary relative to a meridan influences both the width and the intensity of a western-boundary current. Further improvements were effected by Charney (1955) and Morgan (1956).

Westward intensification of currents also takes place in abyssal waters (see Stommel, 1957; Robinson & Stommel, 1959). For the most part, however, upper and lower flows are in opposite directions (Fig. 2.22). Water near the bottom of the ocean moves very slowly, and even in western boundary currents flow-rates probably do not much exceed 30 cm sec^{-1}.

The significance of the Gulf Stream System

Two centuries ago interest in the Gulf Stream was primarily navigational. In those days methods of finding longitude were somewhat primitive and, as Maury (1858) relates: 'Instances are numerous of vessels navigating the Atlantic in those times being 6°, 8°, and even 10° of longitude out of their reckoning in as many days from port.' Indeed, 'vessels from Europe to Boston frequently made New York, and thought the landfall by no means bad'! Benjamin Franklin (1706–90) appears to have been the first to suggest that the sharp temperature contrast between the warm oceanic waters and the cold coastal waters so characteristic of the western edge of the Gulf Stream might be used as a means of giving navigators longitude (see also Williams, 1799). As Maury put it: '. . . this dividing line, especially that on the western side of the stream, seldom changed its position as much in longitude as mariners often erred in their reckoning'.

To Franklin also should be awarded the credit for publishing (in 1770) the first chart of the Gulf Stream System. It was drawn for him by one

Fig. 2.22. The abyssal oceanic circulation, according to Stommel (1958).

Captain Folger, a Nantucket whaler, of whom Franklin, then Postmaster-General of the English colonies in North America, sought the reason for mail-boats taking up to a fortnight longer to cross the North Atlantic Ocean than merchantmen. He was informed that the skippers of the merchantmen allowed for the Gulf Stream when planning courses.

As is noted in Chapter 1, Maury himself deserves credit for his considerable hydrographic achievements, among which were detailed compilations of winds and weather over the North Atlantic Ocean and of currents in the Gulf Stream System. These compilations were of great benefit to shipping, although by then mariners were able to determine longitude by chronometers, and the need for 'thermometric navigation' no longer existed. Emphasis shifted away from the navigational significance of the Gulf Stream System; the effects of the currents upon the climate of Europe were coming to be appreciated. In Maury's words: 'It is the influence of this stream upon climate that makes Erin the "Emerald Isle of the Sea", and that clothes the shores of Albion in evergreen robes; while in the same latitude, on this side, the coasts of Labrador are fast bound in fetters of ice.' Soon it was realized that the Gulf Stream System influences climate

not simply through the proximity of the warm water it conveys but also through its influence on the weather systems which in general approach Europe from the west.

The nature of the Gulf Stream

In order to achieve a scientific understanding of the ocean–atmosphere interactions by means of which the Gulf Stream System extends its influence far beyond the western North Atlantic Ocean it is necessary to understand the nature of the currents themselves. Since the end of the Second World War a great deal of effort has been directed towards this goal, particularly by scientists of the Woods Hole Oceanographic Institution (see Stommel, 1965). We begin by considering the Gulf Stream part of the System.

The Gulf Stream is necessarily superjacent to a strongly baroclinic zone. As Fig. 2.23 shows, this zone is impressively well-defined to a depth of almost 1,000 m. On the eastern flank of the zone lies warm, highly saline Sargasso Sea water, in which there is rather little seasonal variation of temperature, while on the western flank is found considerably colder, less saline water, in which seasonal fluctuations of temperature are

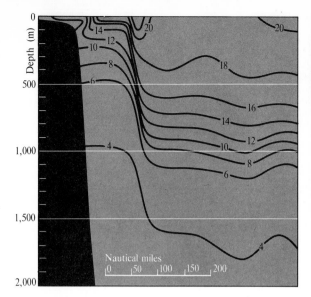

Fig. 2.23. Typical temperature cross-section (°C) through the Gulf Stream and subjacent waters east of Cape Hatteras.

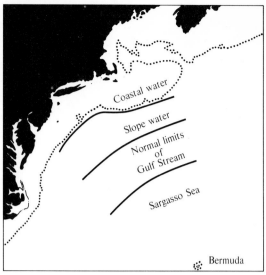

Fig. 2.24. Surface water provinces in the Atlantic Ocean between Bermuda and the coast of North America.

comparatively large. The sharpness of the surface-temperature contrast at the western edge of the Gulf Stream has already been mentioned; according to Groen (1967) it is not uncommon at this so-called *cold wall* for the temperature to drop by as much as 10°C in a distance of 20 km. Between the cold wall and the coast of North America, from Cape Hatteras to the Grand Banks, two water provinces can be distinguished (Fig. 2.24). The region from the edge itself to the 100-fathom line is occupied by *slope water,* in which currents, although often weak and transitory, tend to set counter to the Gulf Stream flow, and the continental shelf is overlain by *coastal water,* in which circulations are dominated by tides, local winds, discharge from rivers and intrusions of oceanic water.

Flow in the Florida Current deviates little from a curvilinear path whose radius of curvature is approximately 2,000 km, but beyond Cape Hatteras conspicuous meanders develop. The amplitude of these increases downstream, from a maximum of about 50 km near Cape Hatteras to an extreme of over 500 km south of Nova Scotia (see Robinson *et al.,* 1974). Fuglister (1963) believes the meridian of 65°W to be a sharp line of demarcation between the part of the Gulf Stream in which meanders are of relatively small amplitude and that in which they are of much larger amplitude. The wavelength of meanders typically varies between about 150 and 400 km.

Detached eddies form east of about 60°W. This was known long before the Second World War, but until the integrated survey of the Gulf Stream by six ships in June 1950 no opportunity arose to observe in detail the formation and progress of such eddies. On this occasion, over a period of several days in the middle of the month, a cyclonic meander was observed intently as it degenerated into a separate cold circulation on the Sargasso Sea side of the main current. The synoptic details have been described fully by Fuglister and Worthington (1951), from whose paper Figs. 2.25 and 2.26 are taken. It is now known that such eddies are a usual feature of the western North Atlantic Ocean south of the Gulf Stream (and indeed of the western North Pacific Ocean also, south of the Kuroshio Current) and that some propagate for great distances across the ocean. Moreover, not only are meanders and eddies characteristic elements of the Gulf Stream but so also are complex persistent filamentary flows, again on the Sargasso Sea flank (see Fig. 2.27).

It is clear, therefore, as Fuglister (1955) has pointed out, that the old concept of the Gulf Stream as a single tortuous current may no longer be considered valid. Only as a component of the *general* oceanic circulation is the Gulf Stream recognizable as a continuous relentless flow from Cape Hatteras to the Grand Banks. The meanders, eddies and filaments are not permanent. Quoting Arx (1962):

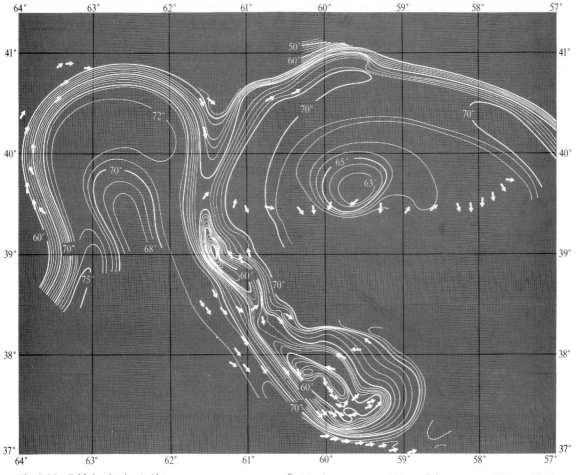

Fig. 2.25. Eddying in the Gulf Stream: mean temperature (°F) in the uppermost 200 m of the ocean on 17 June 1950, as revealed by the researches of Fuglister and Worthington (1951).

they appear and disappear to such an extent that they are lost in the data-averaging processes used to produce mean charts of the circulation. Even when given adequate numbers of data, features that have a characteristic longevity less than the period of averaging will necessarily vanish from the final picture. Average charts therefore have a built-in filter whose working depends on the time span as well as on the spatial interval between sampling points.

Functionally the meanders seem to be essential to the dissipation of the Gulf Stream, but a satisfactory explanation of their cause is still sought. Undoubtedly baroclinic instability is of some importance. Further, it is believed that the Gulf Stream reaches to the bottom of the ocean and that the various seamounts of the western North Atlantic basin may be a deflectional influence (see Warren, 1963). However, theoretical

work by Phillips (1966) suggests that the meanders and eddies are manifestations of the forced response of the *western* part of the ocean to low-frequency changes in the wind-stress over the *entire* ocean. Tropical and extra-tropical cyclones do not modify the fine structure of the Gulf Stream noticeably, but seasonal and climatic fluctuations probably do.

In many respects the Gulf Stream and the Kuroshio Current, its Pacific Ocean counterpart, exhibit similar behavioural patterns. The latter is, however, more prone than the Gulf Stream to the formation of quasi-permanent meanders, especially off Cape Shionomisaki. For a popular introduction to the Kuroshio Current an article by Barkley (1970) should be consulted. The most comprehensive and authoritative work on the subject is probably the monograph edited by Stommel and Yoshida (1972).

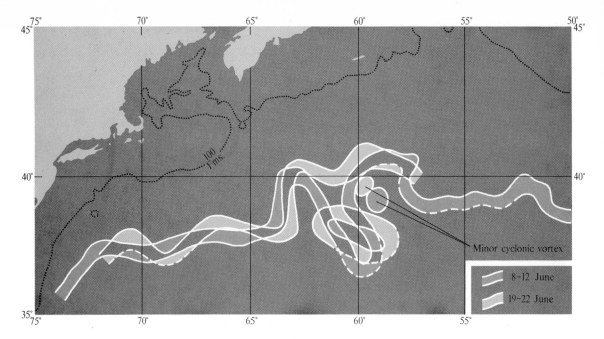

Fig. 2.26. Positions of the warm core of the Gulf Stream during the periods 8—12 and 19—22 June 1950, according to Fuglister and Worthington (1951).

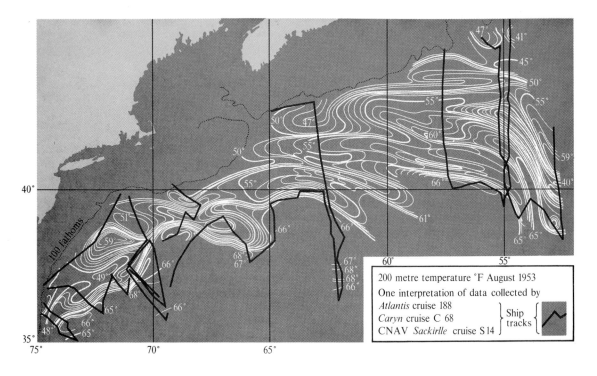

Fig. 2.27. The filamentary temperature structure of the Gulf Stream, as revealed by the researches of Fuglister (1955).

The North Atlantic Current

East of the tail of the Grand Banks the Gulf Stream ceases to be recognizable as a well-defined current. Here is the decay region of the Gulf Stream System, an expanse of ocean in whose surface waters is found a complex pattern of currents and countercurrents. Here the Stream finally degenerates into an assemblage of eddies and fragments. Nevertheless, averaging reveals a net water transport towards Europe, a component of the general oceanic circulation called the *North Atlantic Current*.

This eastward transport is assisted by two processes:

1. To the south and east of Newfoundland the Gulf Stream encounters on its northern flank the very cold, less saline waters of the Labrador Current (Fig. 2.20) and thus becomes, in part, a *gradient current*.

2. The prevailing westerly winds of middle latitudes impart momentum to the surface and thereby fashion waves and create currents. Accordingly, the North Atlantic Current behaves to a considerable extent as a *drift current*.

The origin of the Labrador Current was shown by Munk (1950) to be due in all probability to the wind-stress torque between the westerlies of middle latitudes and the easterlies of higher latitudes. For the reasons outlined earlier a southward-setting western-boundary current is required to balance the tendency for a northward movement of the waters on the poleward side of the North Atlantic Current. The Labrador Current is very cold for two main reasons: first, its source is in Arctic waters far to the north, and, second, it conveys thousands of icebergs. In spring months particularly vast numbers of icebergs, nearly all of them calved from the glaciers of Greenland, reach the Grand Banks. Where the Labrador Current meets the Gulf Stream System marginal eddies develop (Fig. 2.28); some of the cold water mingles with the warm Gulf Stream water and some sinks, helping maintain a sub-surface baroclinic zone. By analogy with the equivalent atmospheric temperature pattern this zone is often regarded as an *oceanic front* [8] .

Many of the eddies and fragments in the decay region of the Gulf Stream System are remarkably persistent as identifiable features and some can be traced for considerable distances across the ocean towards Europe. Associated with them is a complex temperature field, which is of importance

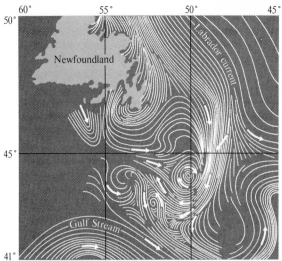

Fig. 2.28. Schematic representation of marginal eddies off Newfoundland, where the waters of the Labrador Current mingle with those of the Gulf Stream.

in the processes of cyclogenesis and frontogenesis, through the agencies of sensible and latent heat transfers from the ocean surface to the overlying atmosphere (see, for example: Petterssen *et al.*, 1962). This subject is explored in Chapter 5.

As it approaches Europe the North Atlantic Current bifurcates (Fig. 2.20). Part turns southward and becomes the Portugal Current. The remainder turns north-eastward past Ireland and the western isles of Scotland towards the Norwegian Sea. Some of this flow branches towards southern Greenland (as the Irminger Current) and a portion enters the North Sea, but the majority of the flow continues on past Norway to the Barents Sea and the eastern part of the Greenland Sea.

The extent to which the warm water conveyed by the North Atlantic Current penetrates high latitudes is evident in the distribution of sea-surface isotherms (Fig. 2.29). Especially noticeable is the zonal temperature contrast in the waters between Norway and Greenland. Indeed, as every schoolboy knows, the coastal waters and fjords of Norway are open to shipping throughout the year, even well to the north of the Arctic Circle, whereas the coast of eastern Greenland is inaccessible to shipping for most of the year, unless icebreaker assistance is available. Near the Lofoten Islands (latitude 68°N) the January temperature surplus, the difference between the average temperature at a particular location and the average for the latitude, is as much as 27°C.

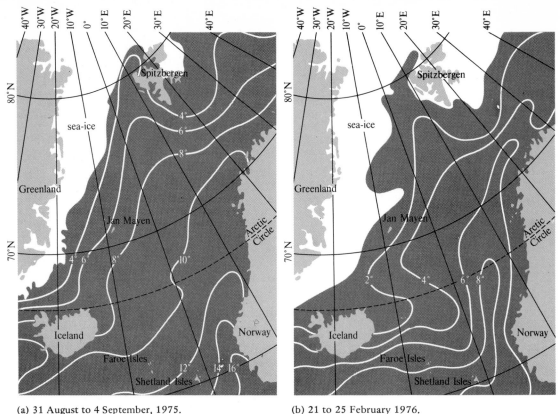

(a) 31 August to 4 September, 1975.

(b) 21 to 25 February 1976.

Fig. 2.29. Sea-surface isotherms and sea-ice limits in the Greenland and Norwegian Seas during the pentads

Over western Europe, from the Iberian Peninsula to Scandinavia, the prevailing westerlies, modified by passage over the warm Atlantic waters, ensure a climate comparatively free from the extremes of temperature experienced by continental interiors.

While the waters of the North Atlantic Current travel poleward their surface layer is cooled progressively by diabatic processes[9] and made less saline by precipitation, run-off and lateral mixing. Where the melt-water which lies adjacent to the polar sea-ice is encountered this layer characteristically bears a temperature of $3°$ or $4°C$ and a salinity of about $35‰$. Its density, which can be obtained simply from a T–S diagram (see Appendix), is therefore near $1,027·8$ kg m^{-3}. Although the melt-water is two or three degrees cooler, its salinity is considerably lower, so it is the Atlantic water which is the more dense. Accordingly, the North Atlantic Current sinks beneath the melt-water and the ice.

The Arctic Ocean

Oceanographically the Arctic Ocean is a marginal

sea of the Atlantic Ocean, because the only passage between the Pacific and Arctic Ocean, the Bering Strait, is too narrow (58 km) and shallow (maximum depth 58 m) to permit more than a very small exchange of water through it, in comparison with the exchange between the Atlantic and Arctic Oceans. Water of Atlantic origin enters the Arctic Basin mainly through the deep channel between Greenland and Spitzbergen and thereupon spreads both eastward, along the edge of the Eurasian continental shelf, and poleward. The heat flux into the Arctic Ocean due to this inflow is about 4 kcal cm^{-2} yr^{-1} (Baker, 1974). Although this is numerically small the heat transported is important for driving oceanic circulations in the Arctic Basin and may influence relative amounts of water and ice there.

Three forms of Arctic sea-ice can be identified:

1. Covering some 6×10^6 km^2 of the Arctic Ocean (Fig. 2.30) is the permanent polar ice-cap, the *Arctic pack,* an expanse of sea-ice mostly several years old. According to Vowinckel and Orvig (1970) 60·9 per cent of ice in the central Arctic Ocean is 5 years or more old and 2 per cent is as

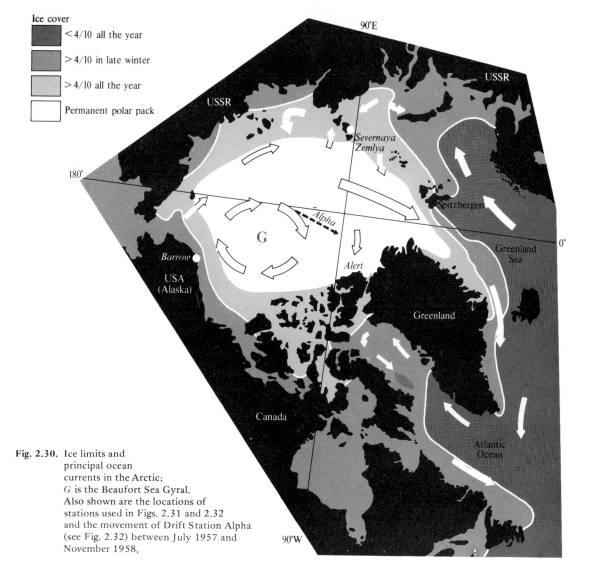

Ice cover

■ < 4/10 all the year

▨ > 4/10 in late winter

▧ > 4/10 all the year

□ Permanent polar pack

Fig. 2.30. Ice limits and
principal ocean
currents in the Arctic;
G is the Beaufort Sea Gyral.
Also shown are the locations of
stations used in Figs. 2.31 and 2.32
and the movement of Drift Station Alpha
(see Fig. 2.32) between July 1957 and
November 1958,

much as 19 years old. The average thickness of the
Arctic pack is 2½ to 4 m at the end of winter and
1½ to 3 m at the end of summer.

2. For most of the year there is attached to Arctic
shores ice which remains fixed in its position of
growth and whose only movement is up and down
with the tides. It is known as *fast-ice*. The limit of
fast-ice coincides approximately with a water
depth of 20 m. In places off Siberia, where water
is shallow to a great distance from the shore, it
extends seaward over 400 km. This is particularly
so off the mouths of the rivers Yana and Lena,
where additionally discharge of river-water
maintains low salinities and hence conditions
favourable for ice-formation. A sheet of fast-ice

which projects more than 2 m above sea-level is
called an *ice-shelf.*

3. Surrounding the Arctic pack is a field of
pack-ice, composed of loosely-connected pieces of
sea-ice called *floes,* whose individual horizontal
extent is, by definition, at least 20 m and may be
more than 10 km (but not more than the visual
range from a ship's masthead). The floes drift
under the action of wind, current and tidal stream.
Drifting among them may be found *ice-islands,*
sections of ice-shelf occasionally as much as 50 m
thick and several hundred square kilometres in
surface area. On some islands scientists have
maintained continuously geophysical and
biological observatories and laboratories for

39

periods of many months (see, for example, Gordienko, 1961).

Although the horizontal extent of sea-ice can be assessed fairly accurately employing satellite scanning techniques (see, for example, Nye & Thomas, 1974), understanding of the mass budget of Arctic sea-ice is still rudimentary (see Koerner, 1973), because systematic observations of ice thicknesses on the Arctic Ocean are available only from Soviet and American ice-island Drift Stations, from submarine cruises and from a few expeditions.

According to Lewis and Weeks (1970), sea-ice covers about $15 \cdot 1 \times 10^6$ km² of the Arctic, North Atlantic and North Pacific Oceans by the end of winter (March—April); this is some 20 to 25 per cent more than at the end of summer (September). The distribution about the North Pole (see Fig. 2.30) is markedly asymmetric. In winter, ice stretches to as low a latitude as $45°$N in the western Atlantic Ocean and off the eastern USSR, whereas seas are open to shipping throughout the year to almost $80°$N off Spitzbergen. Recent changes in the areal extent of Arctic sea-ice have been discussed by Sanderson (1975).

Authorities agree that the mean steady-state thickness of ice on the Arctic Ocean is a little over 3 m (see Vowinckel & Orvig, 1970). In reality, however, level ice is the exception rather than the rule, because constant atmospheric and oceanic motions cause ice-sheets to fracture and floes to drive together into ridges and hummocks (see, especially, Dunbar & Wittman, 1963; Weeks *et al.,* 1971). As hummocks age, continual thawing and refreezing fuses the floes together into massive blocks which may extend 6 or 7 m above and 20 to 30 m below the water surface. Drainage of entrapped brine occurs, so that rounded, very tough ridges of pure ice are eventually produced. The areal extents of open leads and of ridges and hummocks are unknown but it is believed that more than 10 per cent of the total ice cover is ridged or hummocked.

Taking into account ridges, hummocks and leads it seems that the greatest average thickness of ice on the Arctic Ocean occurs in the Beaufort Sea Gyral (Fig. 2.30), where Koerner (1971) found a mean thickness of $3 \cdot 47$ m. Ice tends to converge upon this area, resulting in the formation of a large number of ridges and hummocks (see also Koerner, 1973, and Campbell *et al.*, 1974). In contrast, ice tends to be relatively thin (and leads frequent) along the Eurasian continental shelf line, as a consequence of considerable water turbulence, created by the submarine topography (Vowinckel & Orvig, 1970).

The seasonal rhythm of ice growth and decay in the Arctic has been discussed by Wittmann and Schule (1966). From January to May the sea-ice grows rapidly but at a decreasing rate; there is little snowfall, and skies are generally cloud-free. Melting and disintegration begin in earnest in June. Refreezing commences in September. During November and December temperatures fall sharply, and by the end of the year newly-formed ice typically attains a thickness of about 1 m. Cloudiness is maximum and precipitation amounts are greatest between July and October (Fig. 2.31). The thickness of multi-year ice varies by only a few centimetres between mild and severe winters because the ice and overlying snow provide good thermal insulation. Observations from Drift Stations indicate that the thickness of snow lying upon Arctic Ocean ice may be about 350 to 400 mm when summer melting commences and that the ice is usually snow-free by mid-July. Air temperatures at three stations on the Arctic Ocean coast and at a Drift Station are given in Fig. 2.32. Atmosphere—ice—ocean interactions are considered further in later chapters.

The requisite compensatory outflow of surface water from the Arctic Basin also takes place predominantly between Greenland and Spitzbergen, and this is the route by which sea-ice is exported from the Arctic Ocean. According to Untersteiner (1963) some 3,100 km³ of ice is exported per year; Koerner (1973), however, gives a figure of 5,580 km³ yr^{-1}.

Before proceeding further, i.e. to consider the destination of this ice-laden current, it is necessary to pause for a study of ice conditions in the Southern Hemisphere. The reason for this apparent digression will become evident.

The Southern Ocean

Occupying polar regions of the Southern Hemisphere (Fig. 2.33) is Antarctica, a circumfluous continent (or maybe archipelago) buried beneath an ice-sheet whose surface area is about $13 \cdot 5 \times 10^6$ km² and average thickness approximately 2 km (see Gow, 1965). Slowly, a metre or so per day, the ice creeps plastically outwards from the central plateau (where the ice-surface lies about 3 km above sea-level) towards the Southern Ocean; and more

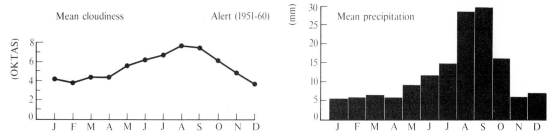

Fig. 2.31. Monthly—mean values of cloudiness and precipitation at Alert.

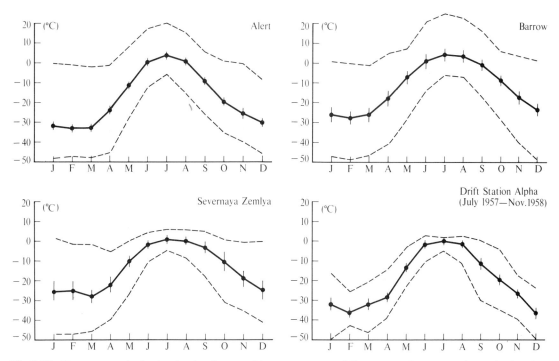

Fig. 2.32. Temperatures in the Arctic, showing monthly averages, mean daily ranges and extreme values. The data for Drift Station Alpha cover only the period July 1957 to November 1958 inclusive.

quickly, some 50 to 100 m per day, great glaciers advance through portals in mountainous regions (see Nye, 1972). For a vivid description of the greatest of these, the Beardmore Glacier, the account of Cherry-Garrard (1922) cannot be surpassed.

Spread over the continental shelf is an ice-shelf, a mass of old sea-ice and superincumbent compressed snow, which in some places attains a height of nearly 100 m above sea-level (see Swithinbank & Zumberge, 1965). It is constantly moving seawards, driven by the force of the inland ice. The Ross Shelf of the Ross Sea and the Filchner—Ronne Shelf of the Weddell Sea (Fig. 2.33) are especially noteworthy for their huge size. Indeed, the Ross Shelf spans the whole width of the Ross Sea, a distance of over 700 km, and

terminates in a cliff, known as the Great Barrier, which rises over 70 m out of the sea. The surface of the Antarctic ice-shelf is remarkably free from undulations, so the icebergs stemming from it are characteristically flat-topped, unlike their Northern Hemisphere counterparts, the more conical bergs derived from the glaciers of Greenland.

Surrounding Antarctica and its ice-shelf is pack-ice (see Heap, 1965) which, according to references quoted by Lewis and Weeks (1970), covers $25 \cdot 5 \times 10^6$ km² at its maximum extent (in September) and exhibits a seasonal variation of about 75 per cent of the maximum. Concentrations of it occur particularly in the Weddell Sea and in the Bellingshausen Sea (Fig. 2.33). Ridges and hummocks are, in general, not as massive as in the

41

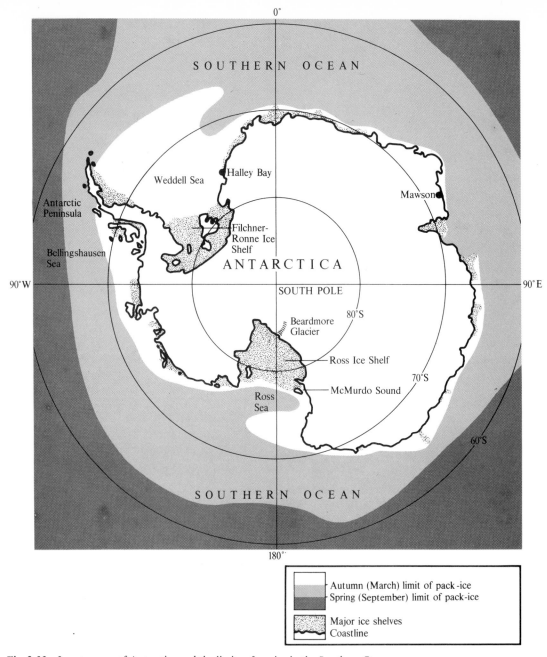

Fig. 2.33. Locator map of Antarctica and the limits of sea-ice in the Southern Ocean.

Arctic, largely because movements of sea-ice in the Southern Ocean tend to be divergent and relatively unrestricted compared with those in the land-locked Arctic Ocean.

On the other hand, Antarctic sea-ice tends to grow to a greater thickness than is usual in the Arctic. For example, Lewis and Weeks mention that first-year ice 2·75 m thick is common at McMurdo Sound (compared with a typical 2 m in

the Arctic) and multi-year ice 15 m thick has been discovered along the western shore of the sound. There are generally more degree-days of cold [10] per year at Antarctic locations than at places of comparable latitude in the Arctic; for instance, at McMurdo Base ($77^\circ 51'$S) there are, Lewis and Weeks note, some 13,300 degree-days of cold per year, whereas at Barrow, Alaska ($71^\circ 18'$N), there are some 8,500. Air temperatures at three stations

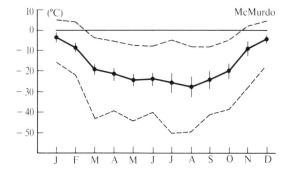

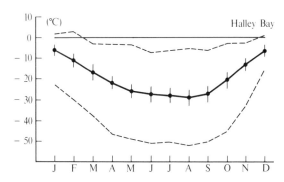

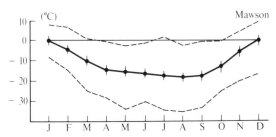

Fig. 2.34. Temperatures on the periphery of Antarctica, showing monthly averages, mean daily ranges and extreme values.

on the periphery of Antarctica are shown in Fig. 2.34.

In the waters closest to Antarctica currents are weak and in many localities westward-setting (Fig. 2.35). Elsewhere in the Southern Ocean there is an eastward flux of water, the *Antarctic Circumpolar Current,* embedded within which are icebergs. More than a few of these are very large (many tens of kilometres in length) and some take as long as 10 years to melt. Bergs may be encountered anywhere south of latitude 55°S, and it is not uncommon to find them as far north as 40°S in the South Atlantic Ocean and the south-western Indian Ocean. The marked departure from zonal flow (diffluence) downstream of the Drake Passage (Fig. 2.35) is responsible for their presence at relatively low latitudes in these regions.

The Antarctic Circumpolar Current is driven by the strong north-westerly winds which prevail between about 40° and 60°S. Since the only substantial land-masses situated within this band of latitudes are the southernmost tip of South America, Tasmania and the South Island of New Zealand, water is able to flow around the hemisphere more or less unimpeded. The current does not run swiftly, mean speeds of up to 15 to 20 cm sec^{-1} being characteristic (maximum near 55°S). Nevertheless, it extends to a depth of more than 3,000 m and, consequently, prodigious quantities of water are transported. According to Kort (1962), Soviet scientists have estimated the mean transport between Antarctica and South Africa to be 190 × 10^6 m^3 sec^{-1}, between Antarctica and Tasmania 180 × 10^6 m^3 sec^{-1}, and through the Drake Passage 150 × 10^6 m^3 sec^{-1} (Fig. 2.36). Kort accounts for the apparent discrepancies between these figures in terms of (*a*) losses due to evaporation and (*b*) exchange of water with adjacent flows, especially the Agulhas and Antarctic Coastal Currents. The Antarctic Circumpolar Current is significantly influenced by bottom topography in several places. In particular, interactions with the New Zealand Plateau and the Pacific–Antarctic Ridge cause large northward deflections of cold water, into the Tasman Sea and the Pacific Ocean respectively (Fig. 2.36).

Although there is nowhere a land-barrier across the Southern Ocean, i.e. a western boundary, the Drake Passage is only 1,000 km wide, and a weakly-developed sub-polar gyre is present in the Weddell Sea (Fig. 2.35). This helps to maintain a permanent tongue of pack-ice across the northern part of the Sea (Fig. 2.33). Drift of ice towards this tongue is further assisted by wind-stress, as Schwerdtfeger (1975) has explained. The mountainous Antarctic Peninsula is a formidable obstacle to the stable air-flows found over cold surfaces. Over the Weddell Sea south of about 70°S winds from an easterly point prevail. When these approach the mountain barrier they are deflected northwards. The westerlies which predominate north of about 64°S help maintain the tongue.

Warm air from lower latitudes is cooled and stabilized as a result of passage over the tongue of pack-ice and the cold surface water with which it is associated. Thus, air temperatures over the Weddell Sea tend to be relatively low, compared with similar latitudes around Antarctica (Table 2.1).

The stabilization aids the development of

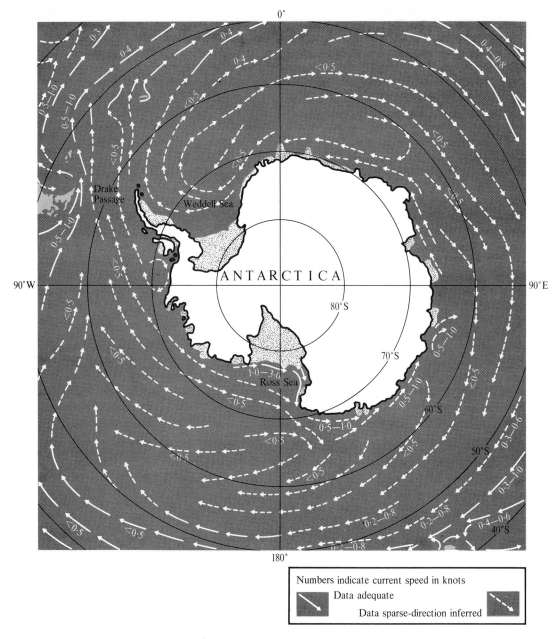

Fig. 2.35. Surface currents in the Antarctic Ocean.

Table 2.1 (from Schwerdtfeger, 1975) Annual average surface air temperature in °C along the parallels 60° and 65°S, between 110°W and 30°E

Long.	110°W	90°W	70°W	50°W	30°W	10°W	10°E	30°E
At 60°S	+2·2	+2·2	+2·6	−2·4	−3·9	−3·6	−3·3	−3·0
At 65°S	−2·8	−2·7	−3·2	−8·6	−8·3	−7·5	−6·8	−6·4

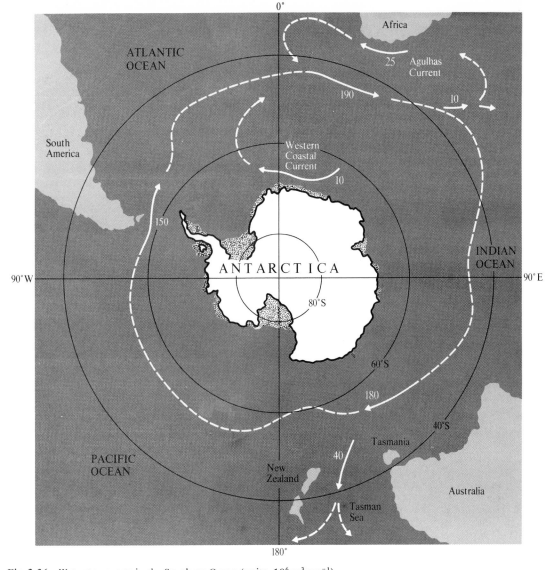

Fig. 2.36. Water transports in the Southern Ocean (units: $10^6 \text{ m}^3 \text{ sec}^{-1}$).

conditions favourable for the northward deflection of winds by the Antarctic Peninsula. In turn, the corresponding tendency for northward advection of cold air is responsible for the lowest mean temperatures occurring over the westernmost Weddell Sea.

The formation of bottom water

Schwerdtfeger ended his paper with the words: 'Altogether, then, the presence of the mountainous Antarctic Peninsula with its effect on temperature and motion of the lower layers of the atmosphere

is an important factor that contributes to make the Weddell Sea the main ice-producing area of the Southern Hemisphere.' He could have added that the Weddell Sea is therefore of considerable importance in respect of abyssal oceanic circulations, for, during winter, the water immediately beneath the sea-ice which totally covers its surface becomes so dense that it sinks down the continental slope and thereupon spreads northwards and eastwards along the ocean floor. At the commencement of sinking its salinity is about $34 \cdot 6\%_0$ and its temperature near the freezing-point of water bearing that salinity,

$-1.9°C$, so its density is about $1,027.9 \text{ kg m}^{-3}$. On the ocean floor it is heated slowly, by means of a small upward flux of geothermal heat (on average about $5 \times 10^{-3} \text{ mW cm}^{-2}$).

It is generally accepted that the Weddell Sea is the major source of this so-called *Antarctic Bottom Water,* and recent studies support the estimate of Stommel and Arons (1960) that the mean annual production-rate there is near $10 \times 10^6 \text{ m}^3 \text{ sec}^{-1}$. Gill (1973) has suggested that bottom water is produced throughout the year, the dense water being supplied from over the continental shelf, particularly in the western part of the Weddell Sea. Other sources of Antarctic Bottom Water identified in the waters adjacent to Antarctica are in the Ross Sea (Jacobs *et al.,* 1970) and off Adelie Land (Gordon & Tchernia, 1972). The mean annual production-rate of Antarctic Bottom Water from all sources is possibly about 38×10^6 $\text{m}^3 \text{ sec}^{-1}$ (Gordon, 1975).

Bottom water is formed likewise in the Arctic Ocean. Water of density $1,028.1 \text{ kg m}^{-3}$ is found at the bottom of the Laurentian Basin, north of Alaska. It is, however, unable to reach the Pacific Ocean on account of the shallowness of the Bering Strait. Water of even greater density, $1,028.15 \text{ kg}$ m^{-3}, is produced beneath the permanent tongue of sea-ice which projects from the Arctic Ocean between Greenland and Spitzbergen, but this *Arctic Bottom Water* is largely confined to the Angara, Greenland and Norwegian Basins. Only very occasionally is it able to cascade over the ridges which connect Scotland, the Faeroe Islands, Iceland and Greenland. Arctic Bottom Water is therefore virtually isolated physically.

In the upper ocean south and south-east of Greenland *North Atlantic Deep Water* is generated. Both the East Greenland Current (the extension of the main outflow from the Arctic Ocean) and the Labrador Current (Fig. 2.20) transport cold polar water to this region, where the water, whose salinity is about $34‰$, mingles with warmer, more saline, water of Gulf Stream origin. The resultant mixture possesses a salinity of some $34.9‰$ and a temperature near $3°C$, corresponding to a density of about $1,027.8 \text{ kg m}^{-3}$. Accordingly, the mixed water is sufficiently dense to sink (see Lee & Ellett, 1967).

Antarctic Bottom Water constitutes the lowermost water mass in the basins of the Indian, Atlantic and Pacific Oceans to well north of the equator. North Atlantic Deep Water, which initially spreads over the floor of the ocean, encounters the more dense water from the Antarctic near $40°N$ in the western part of the ocean and thence flows southwards above it as far as the Southern Ocean. The warm, highly saline, outflow from the Mediterranean Basin (Fig. 2.13) finds its equilibrium depth at about $1,200 \text{ m}$ and flows south with the North Atlantic Deep Water. In the Southern Ocean the mixture of these water-types, known as *South Atlantic Deep Water,* is carried by the Antarctic Circumpolar Current into the southern Indian and the Pacific Oceans.

In the North Pacific Ocean production of bottom water proceeds on a rather small scale, compared with Antarctic and North Atlantic Ocean waters, and only in the Sea of Okhotsk (north of Japan) during winter is there any significant production. The deepest water in the Pacific Ocean is formed by the mixing of South Atlantic Deep and Antarctic Bottom Waters. In general, the water in the Pacific Ocean below about $2,000 \text{ m}$ tends to be homogeneous (with a temperature between $1.5°$ and $2.0°C$ and a salinity between 34.60 and $34.75‰$) and much more weakly stratified than the Atlantic Ocean.

The significance of deep-water circulations

At a depth of several hundred metres a sharp temperature gradient, the *main thermocline,* separates the waters of the upper ocean, in which temperatures are regulated by advection of water masses and by surface diabatic processes, from the cold deep water (Fig. 2.37). This is a stable region, through which vertical motions are strongly resisted. Accordingly, deep water is thermally insulated from the atmosphere, except of course in high latitudes, where the main thermocline is essentially absent.

It is obvious that deep water generated along the borders of Antarctica and elsewhere must return to the upper ocean eventually. Descent of cold water is rapid (several tens of cm sec^{-1}), spatially concentrated (see Stommel, 1958, 1962) and to a great extent seasonally occurrent. Return through the thermocline to the surface layers, in contrast, is largely achieved by a very slow mechanical mixing over a wide range of latitudes and probably takes place at all times of year. The rate of upward mixing through the thermocline is far too slow to be measured by techniques presently available but must, by inference, be in the order of a few metres per year. South Atlantic Deep Water becomes juxtaposed with the more dense (descending) Antarctic Bottom Water in the

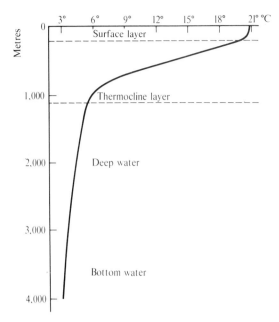

Fig. 2.37. General vertical temperature structure of the ocean.

Southern Ocean and ascends as it is carried along by the Antarctic Circumpolar Current.

It is perhaps not irrelevant to mention at this point that convective circulations of the atmosphere are also markedly asymmetric. Upward motions tend to be rapid and spatially concentrated, whereas subsidence is slow, except in special circumstances (see Ludlam, 1963; Kamburova & Ludlam, 1966). Characteristic time-scales are, however, very much less in the atmosphere than in the ocean. Ascent of air from the lower to the upper troposphere takes 20 to 30 minutes in cumulonimbus convection and 2 or 3 days in extra-tropical depressions, but return to the lower troposphere in anticyclones may take 3 weeks (Green *et al.*, 1966).

Assuming an average water transport of 45×10^6 m^3 sec^{-1}, the period of the thermohaline circulation in the ocean as a whole is approximately 1,000 years. It is likely that the period in the North Atlantic Ocean is nearer 500 years, while that in the North Pacific Ocean may be more than 2,000 years. Historical evidence indicates that there are considerable fluctuations of atmospheric climate over such time-spans. Variations in the extent of sea-ice and of corresponding salinity patterns occur, and there are changes in the positions and intensities of wind-patterns. This being so, the production rate of bottom water cannot be considered annually constant. Consequently, irregular, slow fluctuations must be superimposed upon abyssal circulations.

Inter-relationships between salinity, sea-ice and bottom water are embodied in an elaborate model of climatic change proposed by Weyl (1968); this model is discussed in Chapter 5.

Another notable, but speculative, contribution to the problem of understanding climatic change came from Rossby (1956), who examined the adaptation of the radiation exchange between our planet and space to possible variations of the solar constant and, in so doing, discussed the possible significance of water below the main thermocline as a secular heat reservoir. Commenting upon fluctuations of the Solar Constant over long periods he thought it likely that variations of up to a few per cent of the total energy occur. However, since the majority of this variation is in the ultra-violet part of the solar spectrum, most of which is absorbed by the high atmosphere, the variation remaining to influence tropospheric circulations is evidently rather small.

Rossby went on to consider the storage of heat, pointing out that the Earth's crust is unimportant, on account of its low thermal conductivity, and also that the ability of the atmosphere to store heat is rather limited. This he illustrated by calculating that the mean temperature of the atmosphere would increase by about 6·3°C if during 1 year as much as 1 per cent of the total effective solar radiation, i.e. 0·2 mW cm^{-2}, were stored instead of being returned to space. As the capacity of the atmosphere to absorb water vapour from the sea would simultaneously increase he estimated that the actual temperature rise would probably be about 3°C. The global atmosphere adapts to changes of solar input by adjusting the planetary albedo, i.e. the average cloudiness.

Storage of 1 per cent of the effective insolation raises the temperature of the upper ocean, the vertically-homogeneous mixed layer, through only a few tenths of a degree. This, Rossby showed, is unimportant to the global heat balance. The temperature change brought about by storing the 1 per cent in a layer 1,000 m thick in the interior of the ocean is even smaller, about 0·015°C. However, there is a major difference between the upper ocean and the abyssal waters. He put it thus: 'These deeper layers are insulated from the atmosphere by stably stratified warmer water masses near to the sea surface and are not able directly to restore the radiation balance by means of an increased evaporation and cloud formation.'

47

Rossby recognized that his estimates were far from precise but felt that he could, nevertheless, make two inferences. First, an exact global radiation balance in all probability does not exist, even if periods of several decades are taken into account. Second, anomalies of heat may be stored and for long periods isolated within the ocean. After several decades or a few centuries have elapsed they return to the upper ocean, where the heat is able once more to participate in atmospheric processes. He was aware that the stored anomalies would when they returned to the surface be characterized by tiny deviations of temperature from the mean, and how such minute temperature variations could significantly influence the atmosphere he was unable to explain. He did, however, suggest that the intensity of upward mixing through the main thermocline, a function of the temperature contrast between the upper ocean and the deep water, i.e. of the vertical stability, may undergo rather irregular, slow fluctuations.

The magnitude of the thermohaline contribution to the general oceanic circulation

Clearly, oceanic circulations are driven partly by downward transfer of momentum from winds in the overlying atmosphere and partly by thermohaline processes. The relative magnitudes of the wind-driven and the thermohaline contributions have been investigated by Wyrtki (1961).

In the introduction to his paper Wyrtki recalled that in the early days of scientific oceanography the presence of very cold water in the depths of the ocean was assumed to be a consequence of convectively-induced descent of water which had been refrigerated in polar regions. Moreover, the view that the whole oceanic circulation is produced fundamentally by cooling in high and heating in lower latitudes gained credence from work by Ekman (1905), who showed theoretically that wind-driven currents occupy only the uppermost 200 m or so of the sea. However, when it became clear from the theoretical approach that the principal features of the surface oceanic circulation can be explained by employing wind as the sole driving force, thermohaline processes became relegated in importance and comparatively neglected. On the other hand, the theory which seemed to indicate that the oceanic circulation is

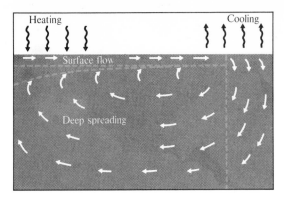

Fig. 2.38. Wyrtki's (1961) two-layer model of oceanic circulation.

essentially wind-driven failed to explain both the observed distribution of temperature and density in deep water and the calculated water transports in specific sections of the circulation. Wyrtki did not dispute that winds to a large extent determine the arrangement of surface currents, nor did he doubt that vertical motions of far-reaching importance may be related to wind-induced convergences and divergencies; he was simply concerned that the magnitude of the thermohaline element be sufficiently appreciated.

Wyrtki chose to consider numerically a two-layer meridional thermohaline circulation comprising: heating of the upper ocean and poleward flow of surface water; subsidence of the most dense water in high latitudes; equatorward spreading of abyssal water; and ascent through the main thermocline (Fig. 2.38). From a study of the heat balance of the upper ocean he deduced an average rate of ascent of between 1 and 5×10^{-5} cm sec^{-1}, but the resultant heat flow through the thermocline he concluded is downwards, in consequence of heat transfer due to thermal conduction exceeding that due to upward mixing. His work indicated further that in a frictional ocean entirely covering the Earth there would be a circumpolar current of strength 134×10^{6} m^{3} sec^{-1} in each hemisphere and a meridional flow of only 10×10^{6} m^{3} sec^{-1} across the parallel of $45°$. Hence, in the actual ocean the thermohaline circulation must be weak and its strength less than that of the wind-driven circulation. Wyrtki found also that the two-layer model could not account satisfactorily for the meridional spreading of the different water masses identified in the deep ocean, but acceptable agreement between theory and reality was obtained with a four-layer model.

Notes

1. Obtained by integrating Planck's formula with respect to λ between the limits 0 and ∞.
2. In a baroclinic atmosphere or ocean, surfaces of pressure and density intersect at some level or levels.
3. Pressure in the ocean is most conveniently measured in decibars (1 dbar = 10^{-1} bar = 10^4 Nm^{-2} = 10^5 dyne cm^{-2}). The pressure at a depth of 1 m is roughly 1 dbar. Since the acceleration due to gravity varies from place to place the exact equivalence of pressure to depth also varies. To eliminate this problem depths can be referred to a work unit of 10^5 dyne cm, called the dynamic metre. Approximately, 1 dynamic metre = 1·02 geometric metre.
4. The Coriolis parameter, usually denoted f, is given by the equation $f = 2\Omega \sin \phi$ where Ω is the Earth's angular velocity and ϕ the latitude.
5. Vorticity is equivalent to angular velocity.
6. Using the Cartesian system of co-ordinates, in which the xy plane is horizontal, x and y being positive eastward and northward respectively, and z is positive upwards vertically,

$$\text{curl}_z \ \tau = \frac{\partial \tau_y}{\partial x} - \frac{\partial \tau_x}{\partial y},$$

where τ is the applied wind-stress.
7. Wind-stress curl in relation to currents in the Arabian Sea has been discussed by Hantel (1970, 1971, 1972).
8. For a discussion of oceanic fronts see Amos *et al.* (1972).
9. A diabatic process is one in which heat enters or leaves a system. Examples are evaporation, emission and absorption of radiation, and turbulent exchange of heat.
10. The number of degree-days of cold is $\frac{\Sigma N\Delta}{24}$, where N is the number of hours during which the air temperature is continuously below 0°C and Δ is the mean temperature during those hours (disregarding the negative sign).

Chapter 3

The action of wind on the sea

Physical processes operating in the lower troposphere and the upper ocean are discussed in some detail in this chapter and the next. The choice of topic with which to commence is inevitably somewhat arbitrary, because the atmosphere and the oceans are intimately coupled and reciprocally interactive. Nevertheless, energy transfers between the two media are predominantly unidirectional, in that kinetic energy is transferred in significant quantities only from air to sea and fluxes of water vapour and sensible heat are for the most part directed from sea to air. The choice is thus simplified. Since there is in Chapter 2 some emphasis on wind-driven oceanic circulations it seems appropriate to devote this chapter to the action of wind on the sea and the next chapter to processes by which energy is transferred from the sea to the air.

Of the momentum transferred downwards from the air to the sea, only a small proportion is utilized directly in creating sea currents; the greater proportion serves to create surface gravity-waves (see Stewart, 1974). How much momentum is communicated from sea to air, or from air to sea, is, however, difficult to ascertain. The rapidly fluctuating motions within the essentially turbulent atmospheric boundary layer must be described statistically (see Lumley & Panofsky, 1964; Pasquill, 1972; Busch, 1973). Further, the extremely sensitive instruments required to measure these motions tend to be prone to inertial errors. They are also delicate and liable to be damaged by waves. Indeed, measurement of air motions close to large waves is such a formidable task that field studies of turbulence over the sea have, to date, proved impracticable in the presence of waves whose height (trough to crest) is more than about 3 m (waves of this height correspond to a sustained wind speed over open water of about 12 m sec^{-1}). Doppler radar techniques described by Bryant and Browning (1975) seem promising, however.

It is possible in boisterous, as in quieter, conditions to employ indirect methods for estimating momentum fluxes — for example, onshore accumulation of water can be related to steady-state wind-stress on the surface of the sea — but interpretation of observations made in such studies is fraught with difficulty (see Roll, 1965).

The many interpretative, instrumental and operational problems inherent in empirical investigations of the atmospheric boundary layer have, on the whole, proved intractable, and estimates of the proportion of the air—sea momentum flux given to generation of surface waves are still open to dispute. Experiments carried out in 1969 during the Joint North Sea Wave Project (JONSWAP) brought about a significant advance in the understanding of momentum transfer from winds to surface waves (see Hasselmann *et al.*, 1973), but, for all that, the present state of this understanding remains far from satisfactory. Of the mechanisms by which momentum is transferred from winds or waves to currents there is almost total ignorance.

The generation and growth of waves

Students of natural philosophy and of mathematics have long sought to understand what Kraus (1972) has called 'a deceptively simple-looking phenomenon', the generation of gravity-waves on the surface of the sea, but, as Hasselmann *et al.* (1973) have noted, 'critical tests of theoretical concepts and numerical forecasting methods have been severely limited by the lack of detailed field studies of wave growth and decay', and a model embracing all the characteristics of waves has proved elusive.

Aristotle (384—322 BC) realized that wind acting on the surface of the sea is of central importance in the development of waves and Pliny (AD 23—79) observed that oil poured upon waves calms them, but thereafter until the middle of the eighteenth century knowledge of the most obvious manifestation of air—sea interaction progressed little. By the latter half of the eighteenth century it was almost universally agreed that wind causes waves, and basic details of how it does so were coming to be recognized. The versatile and perceptive Benjamin Franklin recorded in 1774, for example, that 'air in motion, which is wind, in passing over the smooth surface of the water, may rub, as it were, upon that surface, and raise it into wrinkles, which, if the wind continues, are the elements of future waves' (Franklin *et al.,* 1774). Some fantastic alternative notions were proposed, such as the fermentation theory of Hales (1758), but few took them seriously. Franklin also confirmed, experimentally, that oil diminishes the intensity of waves and suggested that 'the wind blowing over water thus covered with a film of oil cannot easily catch upon it, so as to raise the first wrinkles, but slides over it, and leaves it as smooth as it finds it'.

The atmospheric boundary layer is almost always turbulent, even when to the eye the underlying surface appears smooth, and it is now appreciated that wrinkles (or, as scientists prefer to call them, *perturbations*) can be raised on an initially flat water surface by the action of random pressure fluctuations associated with eddies in a contiguous turbulent air-flow (see Eckart, 1953). In turn, the perturbations themselves disturb the air-flow and thus condition, systematically, boundary-layer wind and pressure patterns (see, for example, Snyder, 1974).

Wavelets spread in all directions, but only those travelling in the same direction as the mean wind become amplified. According to Jeffreys (1925), this happens because

. . . air blowing over the waves may be unable to follow the deformed surface of the water. Water flowing past a sphere does not in general flow all round it; the particles that strike the front of the sphere leave it soon after they have passed the centre, and the region behind the sphere is occupied by eddying liquid with little or no systematic motion relative to the sphere. By analogy one may suggest that if waves are once formed on water, the main air current, instead of flowing steadily down into the troughs and over the crests, merely slides over each crest and impinges on the next wave at some point intermediate between the trough and the crest. The region sheltered from the main air current contains an eddy with a horizontal axis, while smaller eddies exist along the boundary between

this eddy and the main current. If such a theory is correct, the pressure of the air will be greater on the slopes facing the wind than on those away from it; for the deflexion of the air upwards when it strikes the exposed slopes implies a reaction between the air and the water.

In his quantitative treatment of this hypothesis, Jeffreys considered the energy balance of a single regular sinusoidal wave[1] moving with constant speed (c) in the direction of the wind. He disregarded tangential wind-stress and assumed that the only important mechanism for transferring energy from wind to water is the difference of *normal* pressure between the windward and leeward faces of each wave crest. Hence, if that assumption is valid, waves can grow only if the normal energy flux integrated over the water surface exceeds the rate at which energy is dissipated by molecular viscosity. Jeffreys found the criterion for wave growth to be represented by the expression:

$$s\rho' (U - c)^2 c > 4 \mu g \qquad \qquad \ldots [3.1]$$

where U is the wind-speed ($0 \leqslant c < U$), μ the kinematic viscosity, ρ' the ratio of air density to water density, g the acceleration due to gravity and s a dimensionless constant of proportionality which he called the *sheltering coefficient.* Kinsman (1965) thinks 'a more descriptive name would be the *streamlining coefficient,* since it is a measure, in some sense, of the resistance of the wave form to the air flow'.

For the expression to be of any use the sheltering coefficient must be known. Jeffreys noted that the left-hand side of eqn [3.1] is maximum when $U = 3c$. This being so, the least wind capable of raising waves is:

$$U_{\text{min}} = 3\left(\frac{\mu g}{s\rho'}\right)^{\frac{1}{3}} \qquad \qquad \ldots [3.2]$$

Substitution of $\mu = 1\cdot8 \times 10^{-6} \text{ m}^2 \text{ sec}^{-1}$, $g = 9\cdot81$ m sec^{-2} and $\rho' = 1\cdot29 \times 10^{-3}$ in eqn [3.2] yields for U_{min} a value of $0\cdot73 \; s^{-\frac{1}{3}}$ m sec^{-1}. From observations Jeffreys considered U_{min} to be about $1\cdot10$ m sec^{-1}, whence it follows that s must be about $0\cdot27$. However, Stanton *et al.* (1932) obtained from a wind-tunnel investigation of the pressure distribution over a solid wave profile results which indicate that s may be an order of magnitude less.

Ursell (1956) discussed very critically this and other similar laboratory studies, concluding that the pressure differences postulated by Jeffreys are an order of magnitude too large. He commented that, if this be the case, 'the hypothesis of sheltering in its simplest form is insufficient to explain the growth of waves, and mechanisms involving a drag must be considered'. Further, he thought the implication of theoretical arguments and experimental results to be that $U_{min} = 0$.

Sverdrup and Munk (1947) refined the original approach of Jeffreys by taking into account *tangential* wind-stress (τ) exerted on water. They assumed τ to be given by the equation $\tau = \zeta\, \rho_a\, U^2$, where ζ is a dimensionless resistance coefficient (discussed later in this chapter), ρ_a the air density and U the wind-speed at a height of 8 to 10 m ($U > 5$ m sec^{-1} and $U \sim c$ small). For wave growth the sum of the energy fluxes due to normal and tangential stresses, ξ_N and ξ_T respectively, must exceed the rate at which energy is dissipated by molecular viscosity (ξ_μ). The criterion for wave growth ($\xi_T \pm \xi_N > \xi_\mu$) then becomes (cf. eqn [3.1]):

$$2\zeta\, \rho'\, U^2\, c \pm s\, \rho'\, (U - c)^2\, c > 4\, \mu g \qquad \ldots [3.3]$$

which suggests that waves can grow when $c > U$; appeal to observation confirms that this is so in Nature (see Miles, 1957).

Sverdrup and Munk could not ascertain the energy fluxes very accurately. Nevertheless, they formed the opinion that ξ_T exceeds ξ_N for most of the time waves are growing. Indeed, according to Bretschneider (1966), ξ_N may possess a small negative value when $U \approx c$. Only for a short time during the initial stages of wave development does ξ_N exceed ξ_T. When $\xi_T \pm \xi_N = \xi_\mu$ the sea is said to be 'fully-developed'. The waves have then attained their maximum height and speed for a particular wind-speed and are no longer dependent upon the duration of the wind or upon the distance over which it has blown (the fetch).

Although the approach of Sverdrup and Munk was an improvement upon that of Jeffreys there were still shortcomings. For example, they assumed the drag coefficient to be independent of wave height and the sheltering coefficient to be constant (13×10^{-3}). Darbyshire (1952), on the other hand, supposed the latter coefficient to be proportional to the steepness of the wave (H/λ, where H is the trough to crest height and λ the wavelength). This modification reconciles the values of s given by Jeffreys and by Sverdrup and Munk, but casts doubt upon the experiments of Stanton *et al.*

Further, Sverdrup and Munk based their theoretical argument solely upon the energy balance over waves. This is unrealistic because

molecular viscosity is of too small a magnitude to be responsible alone for the dissipation of energy; atmospheric and oceanic turbulence is also involved (see Kinsman, 1965). Moreover, Sverdrup and Munk considered U to be constant, a defect Phillips (1957) attempted to remedy by incorporating in his theoretical treatment the concept of wind fluctuating rapidly about a mean value and the postulate that the wave field adjusts intimately to the turbulent wind field. It is, however, extremely difficult to model satisfactorily the complex and constantly varying pattern of surface stresses which accompanies the continuously developing, interacting and decaying turbulent eddies, and the so-called *resonance theory* of Phillips, although a notable advance, also failed to replicate all the observed properties of waves.

Theories of how precisely the transfer of energy and momentum from wind to water is accomplished have tended to become increasingly complicated, as reference to, for example, Kinsman (1965) or Hasselmann *et al.* (1973) shows, and for their comprehension considerable skill in mathematics and statistics is required. The opening words of Ursell's (1956) paper nevertheless still apply: 'wind blowing over a water surface generates waves in the water by physical processes which cannot be regarded as known'.

Wave forecasting

Although the challenging task of gaining a total physical insight into the generation of waves has yet to be completed, one important object has been achieved, that of constructing relationships by means of which, given a knowledge of wind speed, duration and fetch, wave characteristics can be estimated sufficiently accurately to be of some practical value to those engaged in marine activities. Underlying these relationships is a mixture of theory and empiricism, for theory alone has so far proved incapable of modelling sea conditions adequately. In practice, useful forecasting relationships have been developed from theoretical foundations by taking into account data obtained from measurements of actual waves. The accuracy of wave forecasts thus obtained cannot, however, be better than is warranted by the quality and quantity of data available (see Draper 1967, 1970a). An entirely physical approach to wave forecasting must remain the ultimate object, but until all problems of wave growth and decay can be resolved analytically — in particular, the estimation of non-linear energy transfers due to interactions between waves, and energy losses associated with breaking waves (see, for example, Hasselmann *et al.*, 1973, and Hasselmann, 1974) — semi-empirical forecasting techniques must suffice. Reliability of the wind forecasts upon which wave forecasts depend is a further factor to be taken into consideration.

An early attempt to relate wind to sea conditions was made by Rear-Admiral Sir Francis Beaufort, whose celebrated Wind Scale, devised in 1805 and adopted by the Royal Navy in 1838, is still used widely today, though in greatly amended form. Beaufort's original classification of wind force in terms of the canvas carried by a fully-rigged frigate was revised in 1874 to acknowledge changes in the rig of warships (see Garbett, 1926) and expanded late in the nineteenth century to embrace particulars of the sail required by fishing smacks. In 1903 a scale of equivalent wind speeds was introduced, based upon the formula $V = 1 \cdot 87 \sqrt{B^3}$, where B is the Beaufort number and V the corresponding speed in miles per hour (see Shaw, 1923). The passing of sail rendered a specification based on canvas requirements impracticable; and in the early twentieth century the custom of judging wind force by the appearance of the sea-surface came to be accepted (the custom was approved by the International Meteorological Organization in 1939). The most recent addition to the Scale is a table linking wave heights with specified wind speeds.

Undoubtedly the Beaufort Wind Scale is of practical value, but its limitations must be borne in mind, as Dury (1970) has stressed. Strictly it applies only to a fully-developed sea well away from land, so care must be exercised when fetch and duration of wind are limited. Further, the appearance of the sea-surface is influenced not only by wind but also by various other factors, including swell[2], tidal streams and other sea currents, precipitation and, in shallow water, interactions between wave motions and the sea-bed.

The model developed by Sverdrup and Munk (1947) represents the first successful attempt at forecasting waves scientifically. A product of war-time necessity, their pioneering and 'truly imaginative piece of work' (the words of Kinsman), despite its aforementioned shortcomings, admirably served the purpose for which it was primarily intended. In 1942, the Allies, desperately

concerned at the loss of human life during landing operations in heavy surf, commissioned these men to devise a means of anticipating reliably the dimensions of waves on beaches. They acquitted themselves well.

Since the Second World War, efforts to perfect methods of forecasting waves have redoubled, largely no doubt in pursuit of intellectual satisfaction but partly at least in response to requirements such as the design of harbour installations (Draper & Tucker, 1971), exploitation of offshore resources (Draper, 1970b) and weather routeing of ships (James, 1957; Evans, 1968). Some of these endeavours are discussed briefly hereinafter. Elucidation of the mathematical and statistical manipulations embodied in modern forecasting techniques is beyond the scope of this book however, and readers wishing to study them should refer to the reviews of, for example, Neumann and Pierson (1957), Kinsman (1965) and Bretschneider (1966). We shall confine our attention to certain rudiments.

A wave-field typically possesses a disordered, or even chaotic, appearance, so statistical methods must be employed to describe its features. Accordingly, wave forecasts are usually made by either the *significant-wave* or the *wave-spectrum* method. Significant-wave height (H_s), a parameter introduced by Sverdrup and Munk (1947), is defined as the average height of the highest one-third of the waves[3], which is a close approximation to the wave height estimated visually by an experienced mariner. A wave-spectrum describes the distribution of wave height or energy with respect to wave period or frequency.

The wave-forecasting relationships derived by Sverdrup and Munk and revised by Bretschneider (1952) were used extensively for many years, even after Munk (1957) himself pronounced the spectral approach of Pierson, Neumann and James (1955) 'a conceptual advance' and recommended that the Sverdrup—Munk method as originally developed be discarded. Certainly Sverdrup, Munk and Bretschneider failed to model satisfactorily wave decay and swell propagation but they achieved some success in respect of wave generation, for the relationships they constructed yield values of H_s which agree well with observation. Nevertheless, Bretschneider (1957) considered that 'at some future date some type of spectrum method will probably replace the significant wave method for forecasting waves,

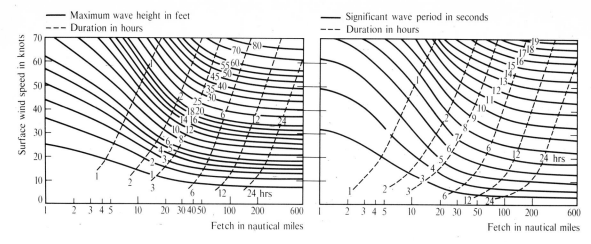

Fig. 3.1. Graph relating oceanic wave height ($H_{\mathrm{max(10\,min)}}$) to wind speed, duration and fetch.

Fig. 3.2. Graph relating oceanic wave period (T_s) to wind speed, duration and fetch.

now generally used for engineering purposes'. This prophecy has not yet been fulfilled, since for estimating H_s engineers still find satisfactory the graphically-presented relationships he produced in a later revision of the Sverdrup–Munk method (Bretschneider, 1970).

In the United Kingdom the empirical wave-forecasting curves published by Darbyshire and Draper (1963) have proved popular among engineers. These curves (see Figs. 3.1 and 3.2), which have been prepared for both oceanic and coastal waters, relate T_s and the most probable value of the height of the highest wave in a period of 10 minutes ($H_{\mathrm{max\,(10\,min)}}$) to wind speed, duration and fetch. Hence H_s can be estimated from the expression $H_s = F \times H_{\mathrm{max\,(10\,min)}}$, where F is a factor depending upon T_s (see Draper, 1967). The most probable value of the height of the highest wave in a period longer than 10 minutes ($H_{\mathrm{max\,(time)}}$) can also be estimated from $H_{\mathrm{max\,(10\,min)}}$, using the expression $H_{\mathrm{max\,(time)}} = G \times H_{\mathrm{max\,(10\,min)}}$, where G is another factor (again see Draper, 1967), if it can be assumed that the sea is fully developed throughout the period (see also Shellard, 1975).

Wave spectra

The significant-wave concept is useful for many engineering purposes but it is insufficient when a knowledge of the distribution of wave energy with respect to wave period is required. For example, it

is desirable when estimating wave forces on offshore structures (see Draper, 1965, and references 50 to 68 in Draper, 1970b) and essential when interpreting scientifically ship motions at sea (see Cartwright, 1958) that the wave-spectrum concept be applied.

The occurrence of a wave-spectrum at a point on the ocean may be explained thus:

Wind systems travel and contemporaneously grow, mature and decay; only briefly can their state be considered steady. In response to change of wind direction, speed, fetch and duration, related patterns of wave height and period evolve. Waves increase in height and period as the wind which raises them strengthens. Further, for a given wind speed, waves also increase in height and period when either fetch or duration (or both) increase (see Figs. 3.1 and 3.2). Waves spread with a phase speed (c) given by the expression:

$$c^2 = \frac{g\lambda}{2\pi} \tanh \frac{2\pi D}{\lambda} \qquad \ldots [3.4]$$

where D is the depth of water. Since $cT = \lambda$, where T is the wave period, $c = gT/2\pi \tanh 2\pi D/cT$. In deep water ($D > \lambda$), $c = gT/2\pi$ and $T^2 = 2\pi\lambda/g$. For a full discussion of the theory of waves see Lamb (1932).

As waves proceed their period remains constant, but their height diminishes gradually. They also interact with each other, with currents and with the wind. To complicate matters still further, there may be present an underlying swell from distant weather systems. Accordingly, a synthesis of mutually-interfering wave trains of differing heights, periods and directions takes place at a particular spot. Superimposed is a wave disturbance consistent with the local wind pattern.

In the opinion of Ursell (1956) attempts to predict entirely theoretically a phenomenon as complex as a wave-spectrum 'must be regarded as almost hopeless'. Recourse to empiricism is necessary. Some theorists have taken up the challenge, however, despite the daunting nature of the task, but so far their efforts have been only marginally successful. Nevertheless, the theoretical contributions of Longuet–Higgins (1952, 1957, 1962) on statistical representations of waves are worthy of mention, for the ideas they contain have proved extremely useful as foundations for realistic models of wave-spectra. Moreover, by deriving a relationship (Longuet–Higgins, 1952) between H_s and the area (E) under a fully-developed spectrum, namely

$$H_s = 2 \cdot 83 \sqrt{E} \qquad \ldots [3.5]$$

he provided a link between the significant-wave and wave-spectrum concepts. The area itself is given by the expression

$$E = \int_0^\infty \left[A(\omega) \right]^2 d\omega \qquad \ldots [3.6]$$

where ω is the wave frequency ($\omega = 2\pi/T$) and A the wave amplitude.

Of the early models of wave-spectra, that developed by Pierson, Neumann and James (1955) appeared to Neumann and Pierson (1957) 'to be the most nearly correct for the widest variety of possible wave and weather situations'. For this reason and also for the clarity of presentation in the forecasting manual they produced, their model was widely adopted (for an outline of it, see Bretschneider, 1966). Its chief contender for acceptance, the model constructed by Darbyshire (1952, 1955, 1956), tended to underestimate the heights of waves in a fully-developed sea and, as was suggested by measurements of glitter from the sea surface (Cox & Munk, 1954)[4], suppress waves of short period.

Taking account of fresh data Darbyshire (1959, 1963) revised his model, but still he failed to satisfy a number of workers. In particular, Neumann and Pierson (1963) thought there appeared to be 'a dichotomy between the results of Darbyshire and the results of other investigators' (notable among whom was Bretschneider, 1959). They went on: 'These other methods [for forecasting waves] all

disagree as to how high the highest possible waves are for a given wind speed, and as to what fetches and durations are needed to achieve waves of a certain height and period. However, they all disagree more sharply with the results of Darbyshire than they do with each other' (see also Walden, 1963). Neumann and Pierson (1963) further pointed out that Darbyshire's model ought to have been superior to those of other workers, since he had the advantage of access to wave records obtained instrumentally at sea over a long period of time, whereas they had to base their models on instrumental records extending over rather limited periods of time or on visual estimates of wave dimensions or (as did Pierson, Neumann & James, 1955) on a mixture of both.

A particular difficulty inherent in Darbyshire's approach seemed to be his treatment of fetch. He used wave records taken by Ocean Weather Ships stationed on the eastern North Atlantic Ocean, and he was careful to select for analysis records which represented a variety of fetches, durations and strengths of wind and which were not significantly influenced by extraneous swell. His work indicated that the sea may be almost fully-developed after a fetch of as little as 100 nautical miles, a figure received with scepticism by many workers. Neumann and Pierson (1963) thought the open ocean where the records were taken 'hardly the place to define fetches of the scale of 100 nautical miles accurately from the synoptic weather observations made in this region'. Phillips (1963), however, proposed conciliatorily that the sea becomes fully-developed rather more rapidly when the water is initially disturbed, as it normally is on the open Atlantic Ocean, than when it is initially at rest.

Cartwright (1961) considered Darbyshire's model adequate for various localities, including the eastern North Atlantic Ocean and the Irish Sea, and commented that the spectra then available appeared to be applicable only to the localities whence were derived the data upon which they were based. Discrepancies between spectra may be due, he suggested, partly to differences in methods of obtaining and analysing data and partly to the influences of certain physical conditions, such as temperature, stability of air flow or tidal streams, none of which is taken into account explicitly in the derivation of spectra. Swell is also an influential factor.

The spectrum model developed by Pierson and Moskowitz (1964) has proved acceptable for some years, but, according to Draper (1974), 'evidence is accumulating that it is too broad to describe a fully-developed sea'. Draper suggested that the model of Hasselmann *et al.* (1973) may be preferable. He felt it probable that the measurements used by Pierson and Moskowitz contained some swell, whereas those embodied in the model of Hasselmann *et al.* were more representative of 'a pure locally-generated spectrum'.

Swell propagation

It is not surprising that the data upon which model wave-spectra are based normally contain a measure of swell, for swell can travel far from the wind-field in which the originative waves are generated. This is well known to mariners, who learned long ago that on tropical oceans swell from an unusual direction can be regarded as an early warning of a hurricane or typhoon. Barber and Ursell (1948) noted that the French swell-prediction service in Morocco (the first authority to make systematic swell forecasts) was able to trace to its origin any outstanding swell that reached the Moroccan coast. Generally the origin was found to be a depression moving across the North Atlantic Ocean between Newfoundland and the British Isles. It was thus clear that swell can travel at least 2,000 km. Barber and Ursell, however, discovered in wave records made off Land's End evidence of swell generated in a storm off Cape Horn, some 10,000 km away; and Munk *et al.* (1963) and Snodgrass *et al.* (1966) showed clearly that swell from storms in the southernmost Indian Ocean can propagate to the Pacific coast of North America, nearly half the circumference of the globe away.

In a storm many wave components are generated. Those of longest period travel fastest and reach a distant observer first. Barber and Ursell concluded that propagation is linear and trains of swell of different length and period behave independently of each other. Accordingly, at a particular point superposed trains from different storms may be present, together with waves generated locally. Frequency analysis of a wave record is necessary to identify and measure the individual trains.

Swell of low frequency is attenuated remarkably little during propagation. The measurements of Snodgrass *et al.* suggest that away from the generation area attenuation is negligible for frequencies below 70 mHz (< 0·02 dB deg^{-1}[5] between New Zealand and Alaska) and only

0·15 dB deg^{-1} at 80 mHz (T = 14·3 and 12·5 sec respectively). In and close to the generation area attenuation is, however, greater; Snodgrass *et al.* estimated that in the near zone of a storm (within a distance comparable to the storm diameter) it is 0·2 dB deg^{-1} at 70 mHz and 0·4 dB deg^{-1} at 80 mHz. In this zone the growth of wave-spectra is limited chiefly by wave-breaking and interactions between waves. Direct coupling between waves and adverse winds, although weak, is also significant, but molecular viscosity and turbulence in the water are apparently negligible (see Phillips, 1969).

Breaking waves

When the wind-speed reaches 4 m sec^{-1} or so (Beaufort Force 3) a few *white horses* (foam-topped waves) begin to appear. As the wind strengthens so the number of white horses increases (see Monahan, 1971). For Beaufort Force 5 (wind-speed 9 or 10 m sec^{-1}), for example, the Sea Criterion is: 'Moderate waves, taking a more pronounced long form; many white horses are formed. (Chance of spray)', and for Force 7 (wind-speed 15 or 16 m sec^{-1}) it is: 'Sea heaps up and white foam from breaking waves begins to be blown in streaks along the direction of the wind'. In a wind of gale force (8) the edges of wave crests begin to break into *spindrift* (scudding spray), and ultimately, when the wind is of hurricane force (12), 'The air is filled with foam and spray. Sea completely white with driving spray; visibility very seriously affected.'

We have already mentioned that energy conversions associated with breaking waves are not easily represented in analytical models of wave growth and decay. Now we investigate the reasons for wave breaking.

The credit for providing a realistic theoretical foundation is due to Stokes (1880), who assumed that a wave breaks when the water particles in its crest advance faster than its profile and thence showed that breaking occurs when the crest angle (α in Fig. 3.3) becomes less than 120°. Michell (1893) found that the condition α = 120° corresponds to a wave-steepness (H/λ) of 1/7. These results can be applied to the three principal sets of circumstances under which waves break (disregarding beaches):

1. When $D \ll \lambda$, eqn [3.4] reduces to $c^2 = gD$, so wave speed in shallow water then depends only on water depth. As a wave travels its period remains

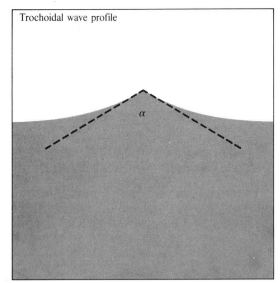

Fig. 3.3. The crest of a simple trochoidal wave and the crest angle.

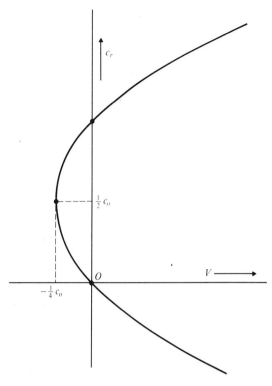

Fig. 3.4. Graphical representation of eqn [3.7].

constant. Thus, if c decreases, so also does λ (because $cT = \lambda$), and the ratio H/λ increases. Furthermore, the rate at which energy is carried forward also remains constant. Since the energy contained in a wave is a function of its height, a decrease of c is compensated for by an increase of H, and H/λ increases still further[6] . Thus, waves commonly break over shoals and submerged reefs.

2. According to Yi-Yuan Yu (1952) the speed (c_r) with which waves travel *relative* to a current of strength V is given by the expression

$$C_r = \frac{1}{2} C_o \left[1 \pm \left(1 + \frac{4V}{C_o} \right)^{1/2} \right], \qquad \dots [3.7]$$

where C_o is the wave-speed in still water. Figure 3.4 shows eqn [3.7] in graphical form. Clearly waves cannot propagate against a current whose strength exceeds $C_o/4$. Indeed, as the speed of an opposing current increases waves become shorter ($C_r^2 = g\lambda_r/2\pi$), and the waves break before the limit $V = -C_o/4$ is reached. For a readable discussion of the behaviour of waves in currents reference should be made to Barber (1969). A detailed account of interactions between storm waves and the Agulhas Current has been given by Mallory (1974).

3. Except in opposing currents the steepness of storm waves in deep water seldom becomes as much as 1/7. The following example illustrates this point. When the sea is fully-developed in a wind of near gale force (say 32 knots) T_s is about 10 sec, H_s about 20 ft (6 m) and $H_{\max (10\,\text{min})}$ about 9 m

(from Figs. 3.1 and 3.2 and the routine of Draper, 1967). Substitution of $T = 10$ sec in the equation $T^2 = 2\pi\lambda/g$ yields the value $\lambda = 156$ m. Hence, $H_s/\lambda \approx 1/26$ and $H_{\max (10\,\text{min})}/\lambda \approx 1/17$. Accordingly, breaking waves are not expected, yet the Sea Criterion for Beaufort Force 7 reads: 'Sea heaps up and white foam from breaking waves begins to be blown in streaks along the direction of the wind'. This paradox arises from failure to consider the complete wave-spectrum. Theory indicates (Longuet—Higgins, 1969) that interaction between waves of different lengths leads to concentration of short-wave energy on the forward side of long waves. Visual observations support this finding, showing that a long steep wave tends to be smooth on its rear slope but rough on its forward face. On this face the length of short waves is reduced by encounter with the opposing orbital motions of water particles in the long wave. Breaking is particularly likely near the crests of long waves (see Wu, 1971).

The meteorological significance of breaking waves

Pockets of air are trapped in breaking waves,

57

whereupon countless small bubbles are formed in the sea. Bubbles are also formed when raindrops strike the sea-surface and when snow-flakes melt in the water. The bubbles are of some importance in the ocean—atmosphere system. They provide a major source of oxygen, carbon dioxide, sulphur dioxide and other gases dissolved in the ocean (see Postma, 1964; Junge, 1972). Further, their bursting at the sea-surface causes droplets of water to be ejected upwards several centimetres into the atmosphere. There, when winds are strong, they are joined by plumes of spray torn from wave crests. In the words of Kraus (1972): 'Bubbles in the water and droplets in the air extend the area of interaction between the two fluids.'

Large droplets do not stay airborne for long and small droplets soon evaporate. However, the salt particles which remain after their aqueous solvent has evaporated (and been wafted away to participate in weather systems) are lifted by turbulence and convection and transported great distances by wind. They participate as condensation nuclei in the hydrological cycle (see Junge, 1963; Roll, 1965; Hobbs, 1971; Mason, 1971) and are eventually returned to the ocean in precipitation and run-off. According to Eriksson (1959) salt is produced over oceanic areas at the rate of about 10^9 tons yr^{-1}, assuming about 0·3 per cent of the global ocean is covered with breaking waves. He considered the residence time of sea-salt particles in the atmosphere to be a few days and estimated that about 90 per cent of the salt is precipitated over oceans. That which falls on land is much longer returning to the sea. Eriksson wrote: 'On land the residence time of sea salt may vary from a few years in peripherally drained, humid areas to thousands of years in arid areas or areas without peripheral run-off.' The significance of storms as means of transporting to land large quantities of salt has been discussed by Tsunogai (1975).

The size-spectrum of bubbles was investigated thoroughly by Blanchard and Woodcock (1957). The largest bubbles they observed were about 1,500 μm in diameter, corresponding to dry salt particles of about 25 μm radius, and the smallest bubbles to reach the ocean surface in white horses were about 100 μm in diameter, corresponding to dry salt particles of about 2 μm radius. Bubbles smaller in diameter than about 300 μm tend to enter into solution. Blanchard and Woodcock found the majority of bubbles to be smaller than 200 μm diameter and the bubble production-rate in the vicinity of a breaking wave to be about

30 cm^{-2} sec^{-1}. They also found that the impact of a raindrop on the water surface produces bubbles of about 50 μm diameter. The larger the raindrop, the greater the number of bubbles and the deeper their penetration. For example, the impact of a small raindrop, about 0·4 mm in diameter, causes the formation of two or three bubbles, which are carried 1—3 mm below the surface, whereas the impact of a large raindrop, about 4·5 mm in diameter, produces several hundred bubbles, some of which penetrate a few centimetres below the surface. The melting of snow-flakes was observed to yield bubbles with a mean diameter of 40 μm. Medwin (1970) found that myriads of even smaller bubbles are formed when continental dust and aerosols[7] strike the water surface.

When a bubble bursts a small jet of water shoots upwards from its cavity (with a speed typically of between 10 and 100 m sec^{-1}). The jet is unstable and disintegrates into some two to ten droplets, each about one-tenth the size of the original air bubble (see Fig. 3.5). These droplets yield sea-salt particles larger than about 1 μm radius. It is believed that the much smaller salt nuclei observed in the atmosphere (some as small as 0·1 μm radius) result from rupture of the liquid film of the bubble into 100—200 tiny droplets (see Mason, 1954; Day, 1964).

Bursting of bubbles at the sea-surface is also important in relation to electrification of the atmosphere, for sea-salt particles produced when the jet collapses and the film of the bubble shatters are carriers of separated electric charge. This is a topic which Blanchard (1963) has discussed in detail.

Contamination of the sea-surface

Some air—sea interaction processes are impaired by pollution of the sea-surface. Kraus (1972) has mentioned that the speeds with which drops are ejected from bubble cavities may be reduced by contamination, and we have already noted that waves are subdued by oil (the formation of capillary waves being inhibited and the attenuation of short waves enhanced). Further, evaporation is retarded in the presence of surface films and skins (see La Mer, 1962).

The surface of the sea is always contaminated naturally, by dust from continents and by films of oil. Seepage through the sea-floor and production by marine animals account for some of the oils, but, according to Dietz and LaFond (1950), the

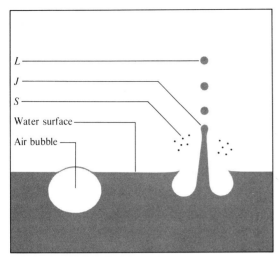

L ——————————————●
J ——————————————●
S ——————————————
Water surface ——————
Air bubble ——————

Fig. 3.5 The bursting of bubbles at the sea-surface. Large drops (*L*) are formed when the jet (*J*) collapses; numerous small particles (*S*) are formed when the film of the bubble ruptures.

majority is probably derived from phytoplankton diatoms. Man has increasingly added to this contamination by using the World Ocean as a sump, and nowadays many tens of millions of tons of a multiplicity of domestic, agricultural and industrial wastes reach the sea each year, either directly or by fall-out from the atmosphere. The annual influx of oil alone might be as much as 5×10^6 tons (see Blumer, 1971; Brummage, 1973).

Since sea-salt particles and evaporation from the ocean surface are of some importance in the hydrological cycle and therefore in the fashioning of climate, it is conceivable that by contaminating the oceans extensively mankind is modifying significantly atmospheric circulations and rhythms (see Mallinger & Mickelson, 1973). Unfortunately, the amounts of such modifications cannot at present be calculated accurately because understanding of atmospheric mechanics is insufficient and knowledge of concentrations, extents and persistence of the various marine pollutants inadequate. However, in general terms, the chief consequence of extensively restraining evaporation from the ocean surface and reducing the availability of sea-salt nuclei in the atmosphere should be a tendency for rainfall amounts to decrease in all latitudes.

Contaminate films reveal themselves as *slicks*, smooth patches or streaks on the surface of the sea. Although a common and widespread phenomenon, natural slicks apparently escaped scientific attention until investigated by Dietz and LaFond (1950). These workers noted the elementary fact

that droplets of oil which are lighter than, and immiscible with, water must rise to the surface, and pointed out that many oils spread into a thin film (normally one molecule thick) because water has a higher surface tension (γ) than any other naturally occurring liquid. Kraus (1972) explained: 'The rim of an oil lens is pulled outwards if the air/water surface tension is larger than the sum of the air/oil and oil/water surface tensions. The spreading of an oil slick depends therefore on the sign of the spreading coefficient

$$Sc = \gamma \,(air/water) - \gamma \,(air/oil) - \gamma \,(oil/water)$$

Spreading occurs if Sc is positive; if it is negative, buoyancy effects due to the density difference between oil and water may still push the rim of the oil patch outward, but the surface tension now pulls it inward.' Kraus then applied the above expression and showed that at $20°C$ olive oil spreads on clean water, whereas paraffin tends to form lenticular globules. Langmuir (1938) found that 1 g of olive oil applied to the surface of a lake spreads out to form a monomolecular layer covering about $10^3 \, m^2$. Otto (1973) has discussed spreading of oil on the sea.

Visual characteristics of natural slicks were described thus by Dietz and LaFond:

Under completely calm conditions, the entire surface of the ocean appears smooth and glassy. Slicks are not readily distinguishable until small wavelets, which require a critical wind speed of about 110 cm sec^{-1} and essentially a zero fetch, are generated. Although slicks can sometimes be distinguished from the surrounding water by a slight difference in texture and colour, they show up primarily when the damping out of small wavelets causes a smooth glassy patch on an otherwise rough surface. When light winds prevail, slicks have a patchy form, but as the winds increase they tend to line up into evenly spaced streaks parallel to the wind. These parallel streaks apparently develop when the slick material collects along small wind-induced convergences associated with the helical circulating cells of water in the homogeneous layer above the thermocline. [See also Cox, 1974.]

The formation of streaks parallel to the wind direction was first studied by Langmuir (1938), whose interest in them stemmed from noticing on the North Atlantic Ocean about 600 miles east of New York large quantities of seaweed arranged in parallel lines. He observed that the lines were typically spaced about 100 to 200 m apart and

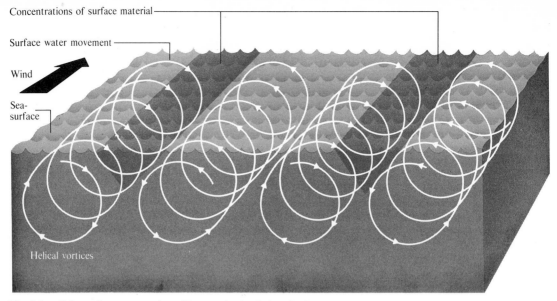

Concentrations of surface material

Surface water movement

Wind

Sea-
surface

Helical vortices

Fig. 3.6. Schematic representation of Langmuir circulations in the upper ocean.

were sometimes as much as 500 m long. It was clear to him from the rapidity with which the alignment of streaks responded to changes of wind-direction that their formation is closely associated with wind-action on the water surface. He suggested that the effect of the wind is to produce a series of helical vortices in the water, with horizontal axes parallel to the wind (Fig. 3.6). Observations have confirmed the existence of these vortices and shown their formation to be due to thermal instability in the surface water. Longitudinal roll vortices are a common occurrence in both the oceanic and the atmospheric boundary layers (see Woodcock, 1940; Woodcock & Wyman, 1947; Kuettner, 1971). Cellular arrangements of convection in the atmosphere are discussed in Chapter 4.

When winds are slight, periodic bands can occasionally be observed on the sea-surface. Their breadth is usually between 10 and 50 m, their length many kilometres, their separation about 300 m, and their speed of movement (along a path perpendicular to their long axes) between 10 and 40 cm sec^{-1} (see Ewing, 1950). The bands have been related by LaFond (1959) to internal waves in a shallow thermocline. In the words of Roll (1965): 'The horizontal components of orbital motion associated with such waves produce periodic convergence and divergence zones at the sea-surface which alternately compress and extend the surface film in a manner sufficient to cause perceptible differences in ripple waves.'

Slicks are dispersed by turbulent water motions in white horses. It should be noted, though, that only the *cohesion* of surface films is overcome by wave action; the sea remains contaminated. When the wind strength decreases slicks reform. A discussion of how the contamination itself is removed is outside the scope of this book, for biological and chemical processes of some complexity are involved.

The origin of drift currents

The energy and momentum of breaking waves are transferred to the underlying current. When short waves break upon a longer wave, an occurrence discussed earlier, the current to which energy and momentum are transferred is the horizontal component of the orbital motion in the longer wave. For most purposes it can be assumed without significant prejudice to accuracy that the orbits of water particles in surface gravity waves are elliptical in shallow water ($D \ll \lambda$) and circular in deep water ($D > \lambda$). Strictly, however, water advances slightly farther in a crest than it retreats in a trough, so a net forward motion is associated with propagation of waves over the sea-surface (Fig. 3.7). In deep water this motion is slow, typically of the order of several millimetres or a few centimetres per second (but substantial nevertheless in comparison with wind-driven ocean currents). In shallow water, particularly on beaches,

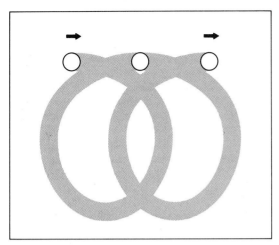

Fig. 3.7. Advanced orbits of water particles in surface gravity waves.

wave-induced currents of some strength can occur locally (undertows and rip and longshore currents are examples of such).

It has long been acknowledged that the direct transfer of momentum from wind to water is a fundamental cause of currents in the upper ocean. Unfortunately, though, the mechanism of this transfer is still not fully understood, so the accuracy with which wind-stress (τ) can be determined by means of the basic relationship $\tau = \zeta \rho_a U^2$ is limited. As with wave forecasting, appeal to empiricism is necessary.

The aerodynamic resistance, or drag, coefficient (ζ)[8] depends upon the height at which the wind-speed (U) is measured. In practice, it is customary for a reference level of 2 or 10 m to be chosen; then, providing the wind is not so strong that streamlining occurs or light that viscous effects are considerable, ζ is independent of U and a function only of the vertical stability of the atmospheric boundary layer and the aerodynamic roughness of the surface over which the wind is blowing (see Sethuraman & Raynor, 1975). Hence, for a particular surface and thermal stability condition, the value of ζ is unique. However, roughness of the sea-surface is determined by the wave conditions prevailing while the wind is acting upon the surface; in turn the wave conditions themselves depend upon the wind situation. Therefore, ζ for a water surface is indeed a function of wind-speed. Moreover, when the sea is not fully-developed it is also a function of wind duration and fetch.

Although there have been numerous experimental investigations and analytical studies of wind-stress upon the sea-surface (see, for example, Kitaigorodskii, 1970), a relationship which describes satisfactorily the variation of ζ with U has not yet been found. Despite increasingly sophisticated instrumentation for measuring ζ, only trends and orders of magnitudes have so far been established with any degree of certainty. In view of the problems involved in empirical investigations of the atmospheric boundary layer over the sea, discrepancies among results are perhaps inevitable.

Wilson's (1960) compilation of the results of forty-seven wind-stress investigations revealed a wide scatter in the values of ζ obtained by field measurements. According to his reckonings, the mean value of ζ at a height of 10 m (ζ_{10}) is for light winds ($U_{10} < 10$ m sec^{-1} and $\overline{U}_{10} \simeq 5$ m sec^{-1}) $1 \cdot 49 \times 10^{-3}$, with a standard deviation of $0 \cdot 83 \times 10^{-3}$, and for strong winds ($U_{10} > 10$ m sec^{-1} and $\overline{U}_{10} \simeq 20$ m sec^{-1}) $2 \cdot 37 \times 10^{-3}$, with a standard deviation of $0 \cdot 56 \times 10^{-3}$. He concluded that ζ_{10} increases with wind-speed, probably in a non-linear manner, until, when $U_{10} \simeq 15$ m sec^{-1}, a value of approximately $2 \cdot 4 \times 10^{-3}$ is approached asymptotically.

Roll (1965) extended Wilson's survey and improved upon it by tabulating and averaging values of ζ_{10}, distinguishing between methods of derivation. He decided that in light winds the method involving measurement of sea-surface tilt is less reliable than other methods commonly employed. These involve: determination of the vertical wind profile immediately above the surface; deviation of the surface wind from the geostrophic direction; and statistical analysis of horizontal and vertical components of wind fluctuations in the turbulent air-flow adjacent to the surface (the eddy correlation method). The four methods and their respective limitations have been discussed in detail in Roll's book and in a paper by Kraus (1968). It is also possible when the wind is light to determine ζ by measuring the extent to which a surface film is contracted as a result of wind stress (see Vines, 1959).

Wu (1969) agreed that for $1 < U_{10} < 15$ m sec^{-1} ζ_{10} increases with wind-speed and proposed the approximate formula $10^3 \zeta_{10} = 0 \cdot 5 \, U_{10}^{1/2}$. He also agreed that for $U_{10} > 15$ m sec^{-1} ζ_{10} is constant and recommended the value $\zeta_{10} = 2 \cdot 6 \times 10^{-3}$ 'for oceanic applications'. On the other hand, his conclusion that a discontinuity in the variation of ζ_{10} with U_{10} exists at $U_{10} = 15$ m sec^{-1} (Fig. 3.8) was somewhat at variance with Wilson's impression that ζ_{10} approaches a constant value asymptotically at $U_{10} \simeq 15$ m sec^{-1}. No such discontinuity was

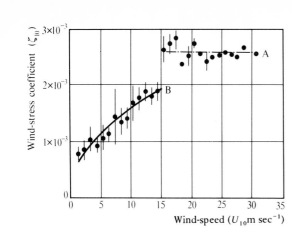

Fig. 3.8. Variation of ζ_{10} with U_{10}; according to Wu (1969). Curve B: $\zeta_{10} = (0\cdot5 \times 10^{-3})U_{10}^{1/2}$. Curve A: $\zeta_{10} = 2\cdot6 \times 10^{-3}$.

evident, however, in the data Smith and Banke (1975) obtained. They, in contrast, found a steady increase of ζ_{10} with U_{10} to the limit of their eddy correlation measurements at $U_{10} = 21$ m sec^{-1}, and suggested, on the basis of 111 data runs, that, for $3 < U_{10} < 21$ m sec^{-1} at least, the relationship $10^3\,\zeta_{10} = 0\cdot63 + 0\cdot066\ U_{10} \pm 0\cdot23$ is appropriate. The quest for an indisputable relationship between ζ and U continues.

The coefficient ζ is a function of atmospheric stability. However, in rather few studies of wind-stress on the sea-surface has particular emphasis been placed upon the influence of stability on the variation of ζ with U. Values of ζ have generally been obtained in near-neutral conditions, although some authors have not stated specifically that this is so.

According to Darbyshire and Darbyshire (1955), thermal stability markedly affects the slope of a water surface tilted by the action of wind-stress, and their measurements (on a lake) indicated that, for a given wind-speed, ζ is considerably greater in unstable than in stable conditions. Their results should be viewed with caution, though, because wind-speed and air temperature were measured at an airport several kilometres from the lake. De Leonibus (1971) showed that ζ can increase *or* decrease with U, depending on the stability of the surface air-flow.

There is some evidence (for example: Darbyshire & Darbyshire, 1955; Deacon, 1962) that ζ increases with fetch, at least for short fetches. The work of Francis (1954), however, suggests that surface drag is largely independent of fetch (and, hence, of wave dimensions) if the fetch is sufficient for short waves to become established. From this it can be

concluded that ripples and wavelets are important elements in the creation and maintenance of drift currents. Indeed, it is now accepted that they account for almost all of the drag component due to waves (known as the *form drag*, as distinct from the other component, the *tangential friction drag*). At this juncture, contamination of the sea-surface and possible climatic change again spring to mind. It is the short waves which are suppressed by films of oil (again see Otto, 1973). Thus, widespread contamination of the surface might conceivably cause a significant diminution in strength of oceanic drift currents and, consequently, a modification of climatic patterns over land areas subject to maritime influences.

Ekman's idealized theory of wind-driven currents

In the late nineteenth century it was fairly generally accepted that wind-drag causes the uppermost layer of water in the sea to flow in the same direction as the wind and that this flow is communicated to deeper layers of the sea by means of internal friction, its direction unchanged but its speed diminishing with increasing depth. The source of the belief lay in a convincing theoretical treatment of drift currents given by Zöppritz (1878). There was, however, a serious flaw in his theory. Although it had long been appreciated that Earth's rotation influences currents in the atmosphere and the ocean (since at least 1735, in fact, when Hadley's classic paper on trade-winds was published), Zöppritz chose to ignore Coriolis force, considering it negligible for ocean currents, which tend to flow rather slowly. The flaw became apparent when it was found that observations of wind and ice-drift taken during the celebrated voyage of the *Fram* across the Arctic Ocean between 1893 and 1896 indicated a deviation of ice-drift some 20 to 40 degrees to the right of the wind-direction. The expedition leader, Fridtjof Nansen, concluded that this deviation must be due to Coriolis force (Nansen, 1902). He concluded further that water must be deviated progressively to the right as depth increases, since moving layers of water set in motion layers beneath (through the agency of internal friction), and once in motion these deeper layers are themselves affected by Coriolis force. The inescapable implication of this conclusion is that there exists a depth at which a particular current flows in the opposite direction to the wind responsible for its generation, unless (through frictional dissipation of energy) its speed decreases to zero before that depth is reached.

The correctness of Nansen's reasoning was confirmed mathematically by Vagn Walfrid Ekman (1905), a gifted young student of the eminent hydrodynamicist Vilhelm Bjerknes, to whom Ekman accorded the credit for being 'the first to give to the laws of motion of the atmosphere and hydrosphere a form clearly indicating the importance of the earth's rotation upon the *forced* currents' (as distinct from the *free* currents, which continue to move by their own inertia after the motive force has been removed − see Stewart, 1967).

Ekman supposed wind-stress on the sea-surface to be the only driving force of surface currents and frictional coupling between adjacent layers of water to be the only means by which water beneath the surface is set in motion, and he simplified the basic equations of fluid motion by making certain assumptions. In particular, he considered an infinitely deep homogeneous ocean, unbounded in the horizontal direction; the wind blowing across it constant in speed and direction; and the currents within it steady[9]. Thus he was able to eliminate from the equations acceleration and horizontal pressure-gradient terms and thence derive straightforwardly relationships for u and v, the horizontal components of the current vector in the x- and y-directions respectively. For convenience Ekman assumed the wind to blow and the wind-stress (τ) to act only in the y-direction.

With reference to the Northern Hemisphere[10] the relationships are:

$$u = V_o \exp\left(-\frac{\pi z}{D_f}\right) \cos\left(45° - \frac{\pi z}{D_f}\right), \quad \ldots [3.8]$$

$$v = V_o \exp\left(-\frac{\pi z}{D_f}\right) \sin\left(45° - \frac{\pi z}{D_f}\right), \quad \ldots [3.9]$$

where V_o is the speed of flow of the uppermost layer of water in the sea and D_f, a depth, is given by

$$D_f^2 = \frac{\pi^2 A_z}{\rho_w \, \Omega \sin \phi} \quad \ldots [3.10]$$

A_z being a coefficient of viscosity in the vertical (z) direction, ρ_w the water density, Ω the angular velocity of Earth's rotation and ϕ the latitude. To distinguish between this depth and the depth referred to in eqn [3.4] the subscript f is used; its significance will become apparent shortly.

By substitution of $z = 0$ in eqns [3.8] and [3.9] Ekman showed that 'the drift current at the very

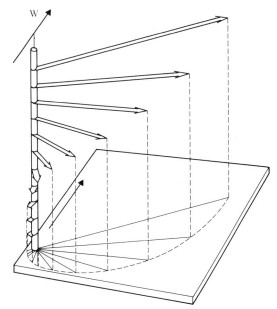

Fig. 3.9. The variation of a wind-driven current with depth.

surface will be directed 45° to the right of the velocity of the wind (relative to the water)'. In the Southern Hemisphere the deviation is 45° to the left. The magnitude of V_o is given by:

$$V_o = \frac{\tau}{\sqrt{2A_z \, \rho_w \, \Omega \sin \phi}} \quad \ldots [3.11]$$

The progressive deviation of a current to the right with increasing depth and the accompanying decrease in its strength are both evident in eqns [3.8] and [3.9]. Projected on a horizontal plane the tips of the current vectors form a logarithmic spiral, the well-known Ekman Spiral (Fig. 3.9).

Ekman was also able to show that the total momentum of a current driven only by the action of wind and experiencing no acceleration 'is directed one right angle to the right of the wind itself'. The reason for this is as follows: Coriolis force operates at right angles to the direction of motion of the centre of gravity of the whole mass of water impelled by wind-action. Since Coriolis force and wind force must be equal in magnitude and opposite in direction when they are the only effective forces, the angle between the wind and the direction of motion of the aforementioned centre of gravity must be 90° (Fig. 3.10).

At a depth of $z = D_f$ the current strength is $V_o \, e^{-\pi}$ (approximately $V_o/23$) and its direction is exactly opposite to that of the surface current. A speed of $V_o \, e^{-\pi}$ is barely perceptible, so D_f can, for all intents and purposes, be regarded as the depth

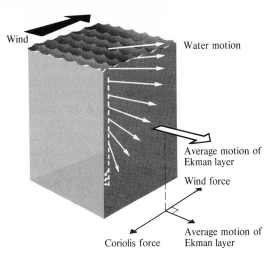

Wind

Water motion

Average motion of
Ekman layer

Wind force

Average motion of
Ekman layer

Coriolis force

Fig. 3.10. Motions in the upper ocean resulting from wind-stress on the water surface. The Ekman layer is the body of water set in motion by wind-action.

of water set in motion by wind-action, or the *depth of frictional influence*. It is clear from eqns [3.10] and [3.11] that D_f and V_o increase as ϕ decreases, other circumstances being equal. They do not, however, become infinite at the equator ($\phi = 0$); certain of the assumptions made in the derivation of eqns [3.8] and [3.9] are not valid when ϕ is small.

Substitution of $A_z = 10^{-3}$ kg m^{-1} sec^{-1} (a typical value for the molecular viscosity of water), $\rho_w = 10^3$ kg m^{-3} and $\Omega = 7{\cdot}29 \times 10^{-5}$ rad sec^{-1} in eqn [3.10], and $\tau = \rho_a \, \zeta \, U^2$ (where $\rho_a = 1{\cdot}2$ kg m^{-3} and $\zeta \sim 2 \times 10^{-3}$) in eqn [3.11], yields the surprising results that in middle latitudes $D_f \simeq 0{\cdot}5$ m and $V_o \simeq U^2/4$. The former is inconsistent with observations, which show that in middle latitudes D_f is characteristically several tens of metres, and the latter is absurd, because it suggests $V_o > U$ when $U > 4$ m sec^{-1}! In practice, V_o is but a few cm sec^{-1} when $U = 4$ m sec^{-1}.

It is, of course, unreasonable to expect such an idealized treatment to replicate precisely the real ocean, but these large discrepancies between theory and observation in fact arise from a conceptual error. When substituting in eqns [3.10] and [3.11] it was assumed that internal friction is due solely to molecular viscosity. This is not so; turbulence in the upper ocean must also be taken into account.

Turbulence in the upper ocean

The turbulence is caused by wind and wave action.

Discrete masses of water are displaced, carrying with them momentum and other fluid properties. They retain their identity for a while, but after travelling a characteristic distance, known as the *mixing length* (a notion somewhat analogous to mean free path in molecular motion), they merge again with their surroundings. The stronger the wind and the greater the intensity of wave-breaking, the farther masses are displaced. A hierarchy of interacting eddies of various sizes is established, the eddy spectrum being bounded by Brownian motion[11] and the dimensions of the fluid system itself. Readers wishing to acquaint themselves with the basic principles of turbulence in the sea will find the review paper of Bowden (1962) useful.

Quasi-random turbulent motions in the upper ocean provide the principal means for the exchange of matter, momentum and energy between the atmosphere and the depths of the ocean beneath the main thermocline. They are essentially macroscopic diffusive motions, in consequence of which a vertically homogeneous layer of water of depth 50 to 100 m is produced (see Fig. 2.37). Although slow — they represent fluctuations of speed and direction superimposed upon an already generally slow-moving mean flow — the disorderly motions are much more effective diffusive agencies than molecular viscosity alone. Without molecular viscosity, however, homogeneity could not be achieved.

In laminar motion A_z represents molecular viscosity; in turbulent motion it represents eddy transference of momentum and is known as the *eddy viscosity coefficient* (Ekman, 1902). Unfortunately, knowledge of turbulent processes in the upper ocean is as yet too uncertain for A_z to be evaluated reliably by the theoretical approach. Appeal to empiricism is necessary. In particular, efforts have been made to estimate A_z by relating the strength of pure drift currents to wind-speed and by considering the energy balance in fully-developed, wind-generated waves.

The value of A_z is not, as Ekman (1905) assumed, constant. In general, it is greatest in the uppermost layers of the sea (for that is where wind and wave action is most vigorous) and decreases with increasing depth. It also increases with increasing wind-speed; according to Neumann and Pierson (1966) its dependence upon U 'for the total energy dissipation in fully-generated, wind-driven wave motion and, probably, for the total energy dissipation in wind-generated currents' can be expressed thus:

$$A_z = 0 \cdot 1825 \times 10^{-4} \, U^{5/2} \text{ g cm}^{-1} \text{ sec}^{-1}, \quad \ldots \, [3.12]$$

where U refers to wind-speed in cm sec^{-1}. This being so, it is apparent from eqn [3.10] that D_f is a function of U and from eqn [3.11] that wind-speed influences V_o not only through τ but also through A_z. For a full discussion of the variations of A_z, D_f and V_o with U the work of Neumann and Pierson (1966) should be consulted.

Eddy viscosity coefficients also vary to some extent with stability and stratification of water masses. At this juncture it is opportune to mention a phenomenon known as *the slippery sea*. When density increases abruptly not far below the sea-surface, a situation which sometimes arises when the surface water is especially warm or fresh, turbulence is suppressed and only the surface layer of water is significantly affected by wind-action. This layer slides almost smoothly over the more dense water beneath (see Woods, 1968a, b; Houghton, 1969).

Some further aspects of Ekman's assumptions

A measure of idealization, and hence of unreality, is unavoidable in a mathematical treatment of fluid motions as complex as those that occur in the ocean—atmosphere system. Ekman (1905) made gross assumptions in his treatment of drift currents. It is not surprising, therefore, that his work was much criticized. There were some who found it hard to believe that surface currents do not flow in the same direction as the wind which generates them and that the angle between wind and current increases with depth. Yet, and perhaps remarkably in view of the simplicity of Ekman's model, the oceanic response to wind predicted by his theory does indeed occur, at least qualitatively, as Ichiye (1965) has clearly demonstrated by dye techniques. Ekman (1905) investigated the validity of his assumption of an infinitely deep ocean. Considering an ocean of depth $z = d$ and retaining all the other assumptions, he showed that the angle (ψ) between the wind and surface-current directions depends upon d/D_f, viz:

$$\tan \psi = \frac{\sinh \dfrac{2\pi d}{D_f} - \sin \dfrac{2\pi d}{D_f}}{\sinh \dfrac{2\pi d}{D_f} + \sin \dfrac{2\pi d}{D_f}} \quad \ldots \, [3.13]$$

Hence: if d/D_f is small, ψ is small; if $d = D_f$, $\psi = 45°$; and if $d > D_f$, ψ is always very close to

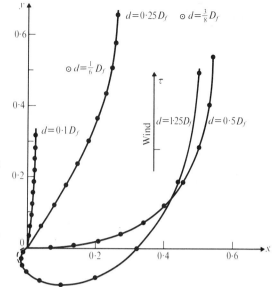

Fig. 3.11. The vertical structure of a drift current in an ocean of finite depth projected on a horizontal plane (after Ekman, 1905). For explanation, see text. The curves for $d = 2 \cdot 5 \, D_f$ and $d = 1 \cdot 25 \, D_f$ coincide, except where dashed near the origin.

$45°$ ($\sinh 2\pi d/D_f$ becomes increasingly large, whereas $\sin 2\pi d/D_f \not> 1$; when $d > D_f$, $\sinh 2\pi d/D_f > 268$). Thus the assumption of infinite depth is responsible for very little error, for only when $d \ll D_f$ is ψ significantly different from $45°$. Figure 3.11 shows projected on a horizontal plane the vertical structure of a drift current for various values of d/D_f. The velocity vectors at different depths can be drawn by connecting to the origin the points marked on the curves. These points are plotted at intervals of $0 \cdot 1 \, d$, from the sea-surface to d. It should be noted from Fig. 3.11 that when $d < D_f$ the angle between the wind and the direction of motion of the centre of gravity of the whole mass of water set in motion by wind action is less than $90°$. Further, when d/D_f is small the influence of bottom friction causes V_o to be much less than in deep water.

By assuming an ocean unbounded in the horizontal direction and the wind-field over it constant in speed and direction Ekman was able to simplify the equations of motion. In reality, though, it is rare for pure drift currents to be present in the sea. Flows of water become impeded by coasts sooner or later, as a consequence of which hydrospheric pressure-gradients develop. Moreover, non-uniformity of a wind-field also leads to convergence or divergence in the upper ocean. For example, trade-winds cause a poleward transport of water, whereas the prevailing westerlies of middle latitudes are responsible for a net equatorward transport (see Fig. 3.12). Accordingly, above the depth of frictional influence in the ocean between these wind regimes convergence occurs. Water tends to accumulate, the isobaric surfaces in it therefore slope and a slow vertical movement of water is induced.

A theoretical discussion of the structure of slope currents will not be attempted in this book. Readers wishing to study the topic should refer in the first instance to Ekman's (1905) own mathematical treatment. For an account of this and subsequent inquiries by Ekman and of work by other investigators the reviews by Neumann and Pierson (1966) and Neumann (1968) are recommended. Mention must be made, nevertheless, of a conclusion reached by Ekman (1905) in respect of friction between the sea-bed and a slope current flowing over it. He showed that in deep water the rate of flow of a bottom current decreases nearly exponentially with increasing depth. At the same time the direction of flow is deviated progressively to the left (in the Northern Hemisphere), until in the layer of water adjacent to the sea-bed the angle between current and horizontal pressure-gradient is $45°$ (see Fig. 3.13). In water unaffected by friction the flow is approximately geostrophic, i.e. parallel to the isobars.

Ekman realized the meteorological relevance of this discovery almost immediately: in the lower troposphere wind must veer with height, until, above the layer affected by friction, flow is quasi-geostrophic. To a first approximation this is indeed the case. However, in the atmospheric friction layer marked horizontal temperature gradients sometimes occur (in association with frontal passage, for example) and eddy viscosity (which is dependent upon thermal stability) often varies with height. Thus, the conditions assumed by Ekman are rarely realized in the atmosphere and the observed variation of wind with height typically differs somewhat from that indicated by the theoretical Ekman spiral.

An undistorted Ekman spiral is also a rare occurrence in the sea, as is no doubt evident from the foregoing. Certainly in shallow water and in ocean basins of limited horizontal extent significant departure from a pure Ekman spiral can be expected. Furthermore, winds not only drive currents but also build waves; and with the propagation of these waves is associated a net

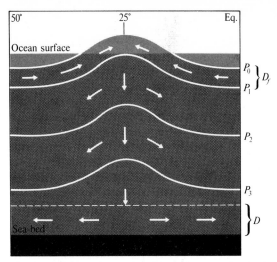

Fig. 3.12. Schematic representation of convergence in the upper ocean near the parallel of $25°$ and associated vertical motions.

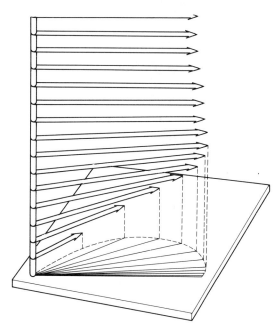

Fig. 3.13. Current vectors near the sea-bed (after Ekman, 1905).

forward motion of water in the direction of the wind. Resultant values of ψ of less than $45°$ are therefore to be anticipated. Observations of winds and currents generally support this statement. Nansen noted that the drift of his vessel among ice-floes on the Arctic Ocean was 20 to 40 degrees to the right of the wind-direction. However, it is unlikely in this instance that waves were responsible for ψ being less than $45°$, since pack-ice greatly attenuates wave motions (see Robin, 1966). There were probably two main contributory factors: first, the Arctic Ocean is of small areal extent ($14 \times 10^6 \, km^2$), and second, since winds in the Arctic tend to be rather variable, currents may not have had sufficient time to become steady.

Efforts to measure drag coefficients of sea-ice surfaces have been described by Banke and S. D. Smith (1973, 1975). Turbulence beneath pack-ice has been investigated by McPhee and J. D. Smith (1975) and Criminale and Spooner (1975).

Upwelling — an application of Ekman's theory

Since the sixteenth century Europeans have been aware that surface water in the eastern Pacific Ocean close to the coasts of Peru and northern Chile is conspicuously cold for the latitudes concerned (see Fig. 3.14). Various theories have been advanced to account for the phenomenon.

Humboldt (1811), whose name was given to the narrow northward-setting current with which the cold water is associated (nowadays generally called the *Peru Coastal Current*), believed advection of frigid water from Antarctica to be responsible for the coldness. This explanation was widely accepted for a time. However, by the middle of the nineteenth century it became clear from observations that water temperature in the Peru Coastal Current can decrease downstream (i.e. equatorwards) for considerable distances, and the advection hypothesis, although not dismissed, was relegated in importance.

In his report on the first major survey of Pacific coastal water off South America, that made from the *William Scoresby* in 1931, Gunther (1936) gave Tessan (1844) the credit for being the first to draw the correct conclusion, that the coldness is largely a manifestation of upwelling, the process by which cold subsurface water ascends to the surface in compensatory response to the removal of water from an area by divergent horizontal flow. Subsequently several superficially plausible mechanisms were put forward to explain the upwelling observed off the west coast of South America. Dinklage (1874), for example, suggested that the widespread westerly drift imparted to the open South Pacific Ocean by the south-easterly trade-winds draws surface water away from the coast, and Witte (1880) proposed that the upwelling is due either to the influence of Coriolis

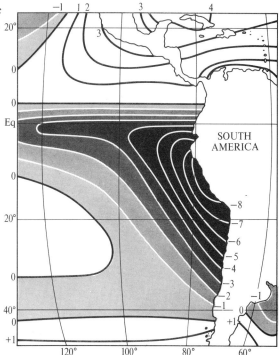

Fig. 3.14. Annual–mean sea-surface temperature anomalies ($°C$), expressed as deviations from the latitudinal average.

force on the northward-setting coastal current or to the stress of offshore winds, but none of these mechanisms proved quantitatively adequate. Indeed, Witte apparently did not know that winds over the coastal waters of Peru and northern Chile tend to be deflected by the Andes so that they blow predominantly parallel to the coast.

Ekman's (1905) theoretical demonstration that a drift current does not flow in the same direction as the wind which sets it in motion provided a framework for the understanding of upwelling. In the Southern Hemisphere drift currents are directed to the left of the wind. Hence, the effect of longshore winds is one of driving surface layers of water away from the coasts of Peru and northern Chile, and subsurface water consequently wells up. The dynamic situation in the ocean then becomes more complicated, for, as a result of the upwelling, temperature gradients are created in the water (see Fig. 3.15) and slope currents are thus initiated.

Although it might be expected that Ekman's theory can provide at best only a rather intuitive insight into the nature of coastal upwelling, in view of the obvious invalidity of most of his basic assumptions, some surprisingly good results have

67

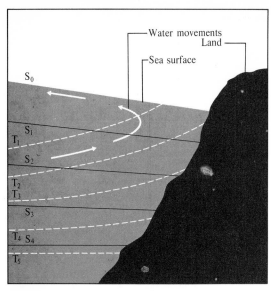

Water movements
Land
Sea surface

S_0
S_1
T_1
S_2
T_2
T_3
S_3
T_4 S_4
T_5

 Slopes
Water movements
Isotherms

Fig. 3.15. Schematic representation of the oceanic temperature structure resulting from upwelling, showing slopes (S) and isotherms (T).

been achieved by researchers applying the theory quantitatively. Notable examples of such work have been reviewed by R. L. Smith (1968). Despite the success obtained with the theory, however, alternative explanations of the coldness of the Peru Coastal Current have been offered from time to time, but they have failed to find favour. For instance, Thoulet (1928) attributed the coldness to melted snow carried down by rivers from the Andes. Gunther (1936) dismissed the idea contemptuously, pointing out that the total volume of river water involved is too small to make much of an impression on coastal temperatures.

Gunther found the appearance of upwelling in the Peru Coastal Current to be temporally and spatially somewhat irregular but certainly related to the wind and broadly consistent with Ekman's theory. No influence by bottom topography was detectable. After considering the temperature and salinity structure of the current in each of twelve latitudes between $35°S$ and $2°S$ and comparing temperatures in the upwelled water with those some distance from the shore (near $100°W$) Gunther concluded that the depth from which upwelling takes place varies greatly between localities. His estimates ranged from 40 to 360 m, with an average of about 130 m. Later studies of the current (see, for example, Wyrtki, 1963)

indicate that the vertical circulation of water participating in upwelling generally extends no deeper than 200 or 300 m and that the average rate of upwelling is typically of the order of 10^{-4} cm sec^{-1}.

The biological effects of upwelling in the Peru Coastal Current are spectacular in scale. As a result of the ascending motions surface waters become rich in nitrates and phosphates. Upon these nutrients phytoplankton and, in turn, zooplankton depend. Fish accordingly abound and millions of birds thrive. So numerous are the birds that their guano is a resource of no little importance in the economies of Ecuador, Peru and Chile. Occasionally, however, upwelling fails, with serious ecological consequences. In response to northerly winds surface water is driven shoreward and the sea becomes abnormally warm. Mass mortality of fish and birds ensues. This anomalous behaviour of the Peru Coastal Current is known as the *El Niño* phenomenon and is so-named because in the years when it occurs — and these are infrequent — it develops soon after Christmas (El Niño is Spanish for 'The Child'). Further reference to El Niño is made in Chapter 5.

Readers may by now have formed the impression that upwelling occurs only off the west coast of South America. This would be a mistake, for, although the upwelling in the Peru Coastal Current is probably more extensive and vigorous than elsewhere, it also occurs strongly, and more or less perpetually, in several other parts of the world, notably in the California, Benguela and Canary Currents. In addition, seasonal upwelling is found in some regions. During northern summer, for example, upwelling is evident between Java and Australia (see Wyrtki, 1962) and off the Somali coast (see Bruce, 1974). Indeed, wherever wind-stress is persistently such that water is transported away from the shore considerable upwelling is to be expected. Those seeking further information on upwelling should refer to the comprehensive review article on the subject by R. L. Smith (1968).

Surges

To complete the present chapter we consider the meteorological factors which may cause the observed tide at a given place and time to differ from the computed astronomical tide. The two factors concerned have been admirably summarized thus by Heaps (1967):

When a depression moves into a sea area the atmospheric pressure acting normal to the sea surface falls and, as a consequence, the sea level rises. As the depression leaves the area the pressure rises and the sea level falls. With the passage of a depression over the sea there is, therefore, a rise followed by a fall in the sea level. Approximately, a change of one millibar in pressure results in a one centimetre change in level. This rule is in accordance with the assumption of a statical law:

$$\Delta H = \kappa \, (\overline{p}_a - p_a),$$

giving the pressure-induced increment in level, ΔH, in terms of the prevailing pressure p_a, the mean pressure $\overline{p}_a$, and $\kappa = 1/\rho_w g$ where ρ_w is the density of the sea water and g the acceleration due to the Earth's gravity. Unless the depression is fast-moving, small, and very intense, as is the case with a tropical cyclone or a squall line, the pressure effect usually gives only a slowly varying contribution to the surge height at a given location. In many cases dramatic changes in surge level around the coast of the British Isles may be attributed to the effects produced by the wind fields of a depression acting over the sea, rather than to the changes in atmospheric pressure associated with that depression. The wind exerts a tractive force on the surface of the water. . . . When motion thus set up in the sea is impeded by a coastline, the water level at the coast tends to either rise or fall — according to whether the net transport of water is towards or away from the land. A raising of level in this way may be identified with a positive surge and a lowering of level with a negative surge. The surge at any location, derived from the observation of the sea level, may therefore be regarded as consisting of a part generated by wind stress acting tangentially over the sea surface, and a part generated by barometric pressure.

The latter influence will not be discussed further, for in shallow seas wind-stress is usually much the more important meteorological factor affecting sea-level, and only on coasts and in shallow water are surges of any practical significance.

To mariners negative surges are a navigational hazard. This is especially so in the shallow waters of the southern North Sea, where traffic of deep-draught ships has grown somewhat in recent years. According to Townsend (1975), tides at the entrance to the Thames estuary can be expected to fall at least 2 feet below predicted levels twenty-five to thirty times a year, and 4 feet or more

below five times a year. 'Evidently', he wrote, ' "cuts" in the tide comparable with the minimum permitted under-keel clearance (3 feet) of the deep-draught ships are far from rare.' Allowance should also be made for the large waves which commonly accompany major surges, because depths are further reduced in wave-troughs.

If sufficiently large, positive surges give rise to flooding. Certain coastal areas are particularly susceptible to inundation by storm surges. For example, on the Atlantic seaboard of the United States, around the Gulf of Mexico, at the head of the Bay of Bengal, and along the Pacific seaboard of Japan, surges raised by tropical cyclonic storms are a recurrent danger, and in various lakes and partially-enclosed seas major surges have been observed during exceptionally stormy weather. References to studies of such events have been given by Heaps (1967). Again, the large waves which form in stormy weather compound the surge problem by acting destructively on coastal defences.

Since time immemorial some coastal areas of the Netherlands and eastern England have been inundated repeatedly (for historical details see Barnes & King, 1953, and Grieve, 1959). Fortunately, surges large enough to bring about catastrophic flooding of these coasts are experienced only a few times each century, and hopefully improved sea defences should nowadays prove capable of containing all but the most extreme surges.

In the United Kingdon much of the impetus for detailed scientific study of storm surges was provided by the disastrous flooding of the Thames on 6–7 January 1928. Investigation of the practicability of forecasting abnormally high water levels in the Thames estuary became thereafter a matter of some priority.

First, it was necessary to identify the meteorological conditions associated with storm surges in the southern North Sea. This was a task Dines (1929) and Doodson (1929) undertook. They found that large positive surges tend to occur when strong northerly or north-westerly winds blow over the major part of the North Sea, whereas negative surges are produced by south-westerlies. Vigorous depressions moving from Scotland to the Baltic can therefore be expected to create in the southern North Sea first a negative and then a positive surge as the accompanying winds veer from south-westerly to north-westerly.

Dines and Doodson further discovered (from

Table 3.1   (after Dines, 1929) Classification of surges at Southend

	Hours before high water						Hours after high water					
	5–6	4–5	3–4	2–3	1–2	0–1	0–1	1–2	2–3	3–4	4–5	5–6
Frequency	0	6	14	6	2	0	1	2	2	1	3	0

analysis of tidal records made at Southend) that large positive surges seem to favour half-tide and avoid high tide (see Table 3.1). This behaviour of surges has since been observed many times. It is apparently due to interaction between surge and astronomical tide (see Keers, 1968; Heaps, 1967). Corkan's (1948) study of large surges experienced between 1928 and 1938 at Dunbar (on the Forth estuary) and at Southend led him to conclude that a storm surge in the southern North Sea may originate partly in the North Sea basin and partly outside. The component which originates *outside* propagates southwards as a progressive wave (see also Keers, 1966), arriving off Southend about 9 hours after it passes Dunbar. The component which originates *inside* Corkan believed to be due to the actions of wind and pressure-gradient on the North Sea as a whole and partly to local effects of these variables.

Thus far, work on North Sea surges had been based on observations from rather few stations. However, in his investigation into the storm surge which occurred in the North Sea on 8 January 1949 Corkan (1950) took into account tidal observations from no fewer than twenty stations in the British Isles and fifteen on the Continent. He was thereby able to study in detail the progress of the surge and show clearly that it travelled counter-clockwise around the North Sea basin, first southwards down the western side of the basin and then northwards up the eastern side, taking about 24 hours to propagate from Aberdeen to Bergen.

An empirical formula Corkan (1948) developed yielded some accurate predictions of storm surges but there were also some failures. On the occasion of the notorious surge of 31 January–1 February 1953, for example, the formula gave, according to Rossiter (1954), 'only fair results' (Fig. 3.16). Rossiter thought that escape of a considerable volume of water through the Strait of Dover, a factor not included in Corkan's formula, might have been one reason for the discrepancies between observed and predicted surge heights.

There have been many accounts and studies of the 1953 surge. This is not surprising, for its magnitude was, and indeed remains, the greatest on record for the North Sea as a whole. At Southend the surge amplitude reached 9 feet (see, for example, Steers, 1953; Grieve, 1959), and at some places in Holland it exceeded 13 feet (see, for example, Ufford, 1953). In eastern England 2×10^5 acres of land were flooded and more than 300 people died. In the Netherlands 1,800 people drowned. To some extent the surge was successfully forecast by the Dutch Surge Warning Service and the British Meteorological Office, in that general warnings of dangerously high water levels were issued several hours before they occurred. Nevertheless, the Committee appointed by the British Government to enquire into the disaster recommended that a Flood Warning Organization be instituted. This recommendation was implemented, the name 'Storm Tide Warning Service' being adopted later. Its Dutch counterpart had been established in 1916.

Since 1953 data on surges have been assiduously collected by the Storm Tide Warning Service and used in the construction of empirical formulae intended for prediction of water levels at various ports on the east coast of Britain. It is a fortunate circumstance that surges propagate southwards along this coast, because in the formula for a particular port residuals[12] at ports farther north can be incorporated. Multiple regression techniques have been used to determine the more significant forecasting parameters at each port and to calculate the coefficients of these parameters in the regression equations. Hunt (1972) has given an example of a prediction formula so derived:

$$R_S = 0.29 + 0.73\,R_L + 0.38\,R_I - 0.57\,R_{L-3} + 0.007\,V_{a,\,t-6}^{330°}$$

in which R_S, R_L and R_I are the high-water residuals (in feet) at Southend, Lowestoft and Immingham respectively (high-water occurs at Lowestoft 3 hours, and at Immingham 7 hours, earlier than at Southend), R_{L-3} is the residual 3 hours before high-water at Lowestoft, and $V_{a,\,t-6}^{330°}$ is the component of the geostrophic wind (in knots) blowing from $30°$ west of north in area 'a' (see Fig. 3.17) 6 hours before high-water at Southend.

Referring to positive surges in the North Sea, Hunt has written thus of the reliability of such formulae: 'The empirical methods have been largely successful, producing, on the whole, quite

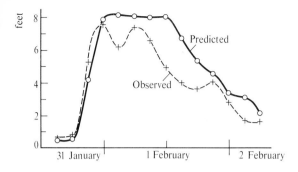

Fig. 3.16. Comparison between observed and predicted tidal disturbances at Southend, 31 January to 2 February 1953.

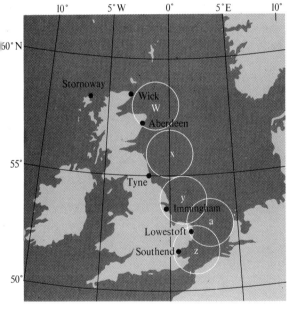

Fig. 3.17. Reference ports and North Sea wind areas used by the British Storm Tide Warning Service when predicting surges.

accurate predictions on the occasions when they have been applied. In fact the errors the formulae give are often of the same order as errors in the astronomical predictions.' Empirical methods similar to those developed for forecasting storm surges on the east coast of England have also proved satisfactory elsewhere, and readers wishing to learn something of techniques employed for predicting surges at Atlantic City, in New York Bay, in Lake Erie, along the Japanese coast, and on the German North Sea coast, for example, should consult references given by Heaps (1967). There have been comparatively few studies of surges on western coasts of the British Isles, and so

far only the meteorological conditions which produce those surges have been identified. The principal investigators of the conditions have been Lennon (1963) and Heaps (1967).

Efforts to forecast negative surges in the North Sea by means of empirical techniques have met with rather limited success (see Townsend, 1975). Forecasting for off-shore areas has proved especially difficult, because regular tidal observations are made only on coasts. Hopefully, however, realistic surge forecasts based upon hydrodynamic theory will soon be available. Results of mathematical models developed to date have been encouraging (see, for example, Heaps, 1969; Finizio *et al.*, 1972; Robinson *et al.*, 1973; Das *et al.*, 1974; Prandle, 1975; Flather & Davies, 1976), but there are, as Hunt (1972) has pointed out, some major problems yet to be overcome. Notable among these is the manner in which meteorological data are introduced into models. Hunt wrote: 'If they are to run operationally, some form of wind or wind-stress field and surface-pressure field will be required as input.' Until the action of wind on the sea is adequately understood, though, this problem cannot be certainly resolved.

Notes

1. In fact, the profile of a wind-generated wave is approximately trochoidal, not sinusoidal.
2. 'Swell' is the name given to waves which are no longer under the influence of the wind which generated them.
3. Similarly, significant-wave period (T_s) is the average period of the highest one-third of the waves; around T_s is concentrated the maximum wave energy.
4. This is a topic treated fully by Jerlov and Steemann Nielson (1974).
5. The decibel (dB) is a unit used for comparing two flux densities. These flux densities (P_1, P_2) differ by N dB when

$$N = 10 \log_{10} \frac{P_1}{P_2}$$

   The distance propagated by swell is measured in degrees of arc.
6. On beaches the situation is rather complicated (see Biesel, 1952).
7. Aerosols were defined by Junge (1963) as: 'dispersed solid or liquid matter in a gaseous medium, in our case, air. The particle size in the atmosphere ranges from clusters of a few molecules to particles of about 20 μm radius, if we disregard cloud, fog, and raindrops, and consider only dry air'.
8. It is usual nowadays for C_D to denote drag coefficient, but ζ is used in this book to avoid possible confusion with c, the symbol used for wave speed.

9. Ekman (1905) found that it takes several pendulum days for the speed and direction of a current to become steady after the sudden onset of a wind (a pendulum day being the time required by a Foucault pendulum to alter its plane of oscillation through $360°$, i.e. one sidereal day divided by sine latitude). However, according to Ekman, the *average* velocity over the first pendulum day is close to the steady velocity.

10. To obtain a solution valid for the Southern Hemisphere the direction of the y-axis must be reversed.

11. In 1827 a botanist, R. Brown (1828), discovered that microscopically small particles held in suspension in a liquid unceasingly move about in an apparently random fashion. This is a manifestation of bombardment by molecules. The phenomenon, which occurs also in gases, has been discussed at length by Champion and Davy (1959).

12. The residual at a point is the difference between actual and predicted tides.

Chapter 4

Ocean-atmosphere heat exchange

The manner in which heat derived from the sun is distributed over the globe by wind-systems and ocean currents is outlined in Chapter 2. The present chapter is devoted to a detailed examination of ocean–atmosphere fluxes of latent and sensible heat, with particular reference to the general circulation of the ocean–atmosphere system and to synoptic-scale and meso-scale weather systems. Climatic change is discussed at length in Chapter 5.

We begin by considering radiant energy.

Radiation patterns

In relation to latitude and time of year the amount of solar radiation falling daily upon a horizontal surface at the fringe of the atmosphere is as shown in Fig. 4.1[1]. Two features of this radiation pattern are specially noteworthy. First, there is an asymmetry between hemispheres. This is due to Earth's distance from the sun varying during the year; perihelion (the point of nearest approach of a planet to the sun) is reached in early January and aphelion (the point in a planet's orbit most remote from the sun) is reached in early July. Second, the daily receipt of solar energy at a solstice is greater in polar regions of the summer hemisphere than at the Tropic of that hemisphere. This is because the longer duration of sunlight at the pole more than compensates for the greater intensity of solar radiation in low latitudes.

When depletion of the solar beam by atmospheric absorption and scattering is taken into account a somewhat different pattern of radiation receipt emerges. This can be seen in Fig. 4.2, a graph showing daily totals of solar radiation arriving on horizontal surfaces at sea-level, assuming the sky to be cloudless and the atmospheric transmission coefficient[2] to be 0·7. As the elevation of the sun decreases, the distance the solar beam travels through the atmosphere increases, and depletion of the beam correspondingly increases. Consequently, maxima

at the solstices now occur only near latitude 35°. According to Deacon (1969), a transmissivity value of 0·7 is appropriate for the moist tropical atmosphere and for regions influenced by dust or industrial haze, and a value of about 0·8 is appropriate for places with relatively clear atmospheric conditions.

Clouds further impede the passage of solar radiation through the atmosphere, so that the actual pattern of radiation receipt at the ground or at sea level displays both meridional and zonal gradations. Figure 4.3 shows that receipt is greatest over land, in arid tracts of the tropics and subtropics, where skies are largely cloudfree. Throughout the middle and upper troposphere over these regions there is widespread subsidence of air participating in Hadley circulations, while in the lower troposphere the vigorous convection which develops by day over the arid ground fails to produce cumuliform clouds because there is little surface moisture to be evaporated for the provision of latent heat and the subsidence inversion prevents thermals from ascending sufficiently far for condensation to take place. Indeed, as Fig. 4.4 shows, insolation receipt in Arizona, Iran, Arabia and the eastern Sahara in the month of June is close to the maximum possible for a dusty locality in the Northern Hemisphere; that is, about 22×10^3 cal cm^{-2} month^{-1}. In contrast, insolation is considerably less than the maximum possible in most oceanic regions, owing to the presence of clouds. Cumulus clouds are characteristic of trade-wind belts; cumulonimbus activity accompanies the ITCZ; and extensive cloud sheets are associated with monsoon circulations and extratropical depressions.

Solar radiation is either absorbed by the sea or reflected from its surface. A minute proportion of that absorbed is fixed by photosynthesis (about 0·02 per cent, according to Williams *et al.*, 1973: see also Strickland, 1958; Isaacs, 1969) and the remainder is used for heating the upper ocean. The degree of reflection is primarily a function of solar

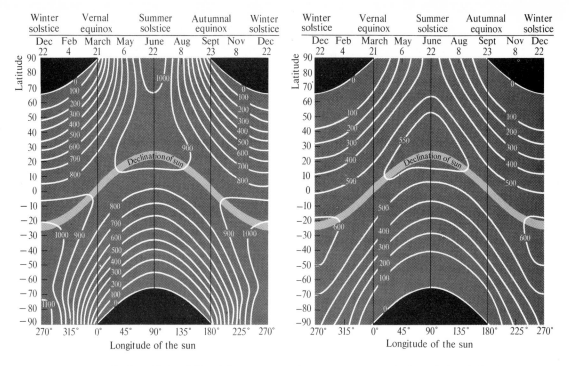

Fig. 4.1. Amount of solar radiation falling daily upon a horizontal surface at the fringe of the atmosphere.

Fig. 4.2. Amount of solar radiation falling daily upon a horizontal surface at sea-level when the atmospheric transmission coefficient is 0·7.

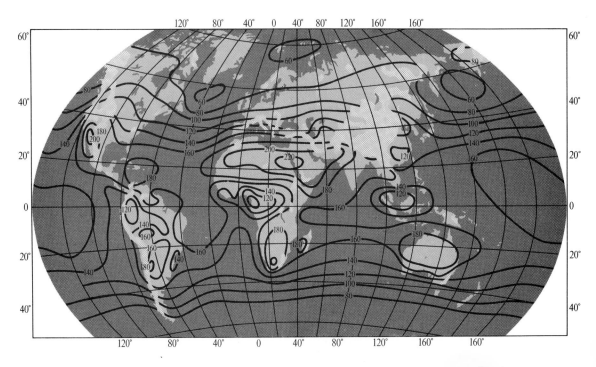

Fig. 4.3. Solar radiation received annually on a horizontal surface at ground level (units: kcal cm^{-2} yr^{-1}).

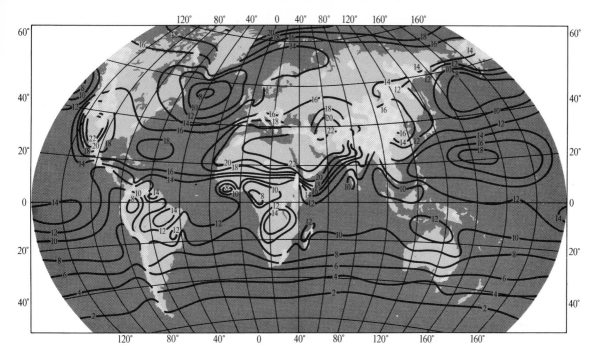

Fig. 4.4. Solar radiation on a horizontal surface at ground level in June (units: kcal cm^{-2} month^{-1}).

Table 4.1 (after Budyko, 1956) Albedo of the sea-surface for total incident solar radiation; that is, both direct and diffuse radiation being taken into account. Values are expressed in per cent.

Lat. (N)	Month Jan.	Feb.	Mar.	April	May	June	July	Aug.	Sept.	Oct.	Nov.	Dec.
70°	—	23	16	11	9	9	9	10	13	15	—	—
60°	20	16	11	8	8	7	8	9	10	14	19	21
50°	16	12	9	7	7	6	7	7	8	11	14	16
40°	11	9	8	7	6	6	6	6	7	8	11	12
30°	9	8	7	6	6	6	6	6	6	7	8	9
20°	7	7	6	6	6	6	6	6	6	6	7	7
10°	6	6	6	6	6	6	6	6	6	6	6	7
0°	6	6	6	6	6	6	6	6	6	6	6	6

elevation and secondarily of sea-state. For direct radiation falling upon a smooth sea, albedo varies from about 3 per cent when the sun is vertically overhead to almost 100 per cent when it is close to the horizon, whereas for diffuse radiation it is typically about 8 to 10 per cent (see Table 4.1 and Kondratyev's (1972) monograph on radiation processes in the atmosphere).

A rather shallow layer of water is heated directly by solar energy. The precise depth to which radiation penetrates at a particular spot depends upon wavelength (see Table 4.2 and Clarke, 1967) and upon concentrations of silt, foam, bubbles and plankton in the water, but in the clearest water only 18 per cent of radiation incident upon the sea-surface reaches a depth of 10 m (Sellers, 1965). In what Groen (1967) calls 'average oceanic water' only 9·5 per cent of incident sunlight reaches a depth of 10 m and 0·31 per cent reaches 50 m. The upper ocean is, however, mixed by means of turbulence, and heat is thus distributed more or less homogeneously through a surface layer of thickness about 100 m.

Not only does the sea absorb and reflect solar energy; it also emits radiation (R_E) of a wavelength appropriate to its temperature and absorbs long-

75

Table 4.2 (after Sellers, 1965) Penetration of solar radiation into pure water in relation to wavelength of the radiation. Values expressed in per cent.

Depth	Wavelength (μm)			
	0·2–0·6	0·6–0·9	0·9–1·2	1·2–3·0
0·00	100·0	100·0	100·0	100·0
0·01 mm	100·0	100·0	100·0	97·2
0·1 mm	100·0	100·0	99·6	79·0
1·0 mm	100·0	99·8	96·2	40·7
1·0 cm	100·0	98·2	68·7	7·6
10·0 cm	99·7	84·8	4·6	0·0
1·0 m	96·8	36·0	0·0	—
10·0 m	72·6	2·6	—	—
100·0 m	5·9	0·0	—	—

wave radiation (R_B) transmitted downwards from clouds and the atmosphere. This being so, the balance of radiant energy (R_N) at the sea-surface can be written:

$$R_N = (1 - A)\, R_S - (R_E - R_B), \qquad \dots [4.1]$$

where A is the surface albedo and R_S the total short-wave radiation arriving at the surface.

Accurate estimation of the terms in eqn [4.1] is not easy because very few systematic measurements of radiation have been made in oceanic regions.

Satellites have helped in this matter (see, for example, Winston, 1969), but the reliability of oceanic radiation data nevertheless still leaves much to be desired.

It is clear from the work of Budyko (1956, 1963) and others that the annual radiation balance of the ocean surface is everywhere positive (Fig. 4.5). The largest balance, a net income of more than 140×10^3 cal cm^{-2} yr^{-1}, is located in the north-western part of the Arabian Sea. In this region there is a modest positive balance in winter (when a trade-wind regime prevails) and an exceptionally large positive balance in summer. At the latter season upwelling due to persistent south-westerly winds causes water off the Arabian coast to be comparatively cool (Fig. 4.6) and, consequently, values of R_E relatively small. Furthermore, cloud amounts are generally small, for two reasons: the cool sea-surface exerts a stabilizing influence on the lower troposphere, and air originally lifted in Asian monsoon cloud systems subsides in the middle and upper troposphere over south-west Asia (see Ramage, 1966; Walker, 1975). Accordingly, daily values of R_S are close to the maximum possible.

Seasonal patterns of radiation balance are shown in Figs. 4.7 and 4.8. The principal features of the winter balance are: (i) the negative balance

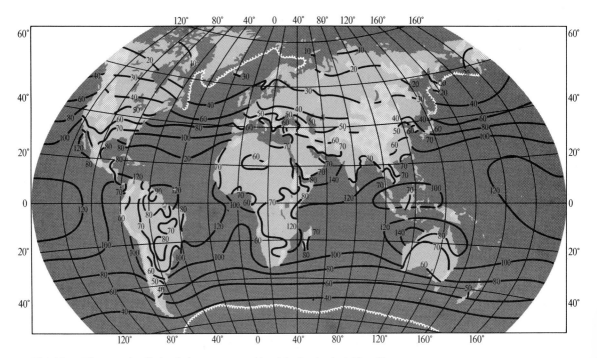

Fig. 4.5. The annual radiation balance at ground level (units: kcal cm^{-2} yr^{-1}).

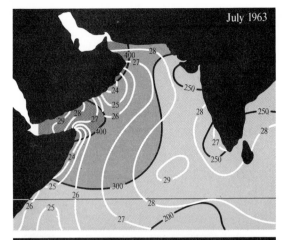

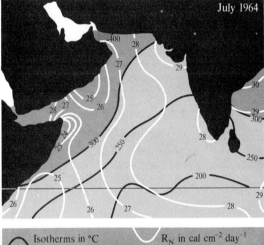

Fig. 4.6. Sea-surface isotherms and the radiation balance of the Arabian Sea, July 1963 and July 1964. Values for the Red Sea and the Persian Gulf have not been included.

polewards of middle latitudes and (ii) the generally zonal orientation of isopleths of R_N. In contrast, considerable deviations from a zonal pattern are evident in summer, the deviations being associated mainly with zonal variations of cloudiness and sea-surface temperature.

Components of the energy budget

The balance of radiant energy on an area of the ocean surface is distributed chiefly between ocean–atmosphere transfers of sensible and latent heat (Q_H and Q_E respectively), storage of heat in a column of the ocean (Q_S) and horizontal divergence of heat by sea-currents (Q_{vo}). The relationship between these major components of the energy budget is:

$$R_N = Q_H + Q_E + Q_S + Q_{vo} \qquad \ldots [4.2]$$

Other processes involving energy are quantitatively insignificant and are, therefore, neglected. Examples are: dissipation of kinetic energy when wind blows across the sea-surface; sustenance of biological processes; and warming of precipitation as it falls.

Until the early part of the present century reliable quantification of the energy budget was precluded by lack of insight into the physical processes involved and by paucity of necessary data (see, for example, the commendable, but rudimentary, attempt of Dines, 1917). As the quantity and quality of atmospheric and oceanic observations improved, knowledge and understanding of the various components advanced steadily, but, for all that, the climatology of the energy budget was not investigated at all comprehensively until Jacobs (1942, 1951) carried out his celebrated researches into energy-exchange between sea and air. Jacobs' work was updated and extended by Budyko (1956, 1963, 1974), and at the time of writing the latter's maps and graphs of energy-budget components are widely regarded as the most authoritative available[3]. Efforts to obtain definitive values of the components nevertheless continue, since a detailed knowledge of the global energy budget is essential if atmospheric and oceanic circulations are to be fully understood. Houghton (1954) and London (1957) were among the pioneers in the quest for this understanding. Subsequent endeavours have been reviewed by Palmén and Newton (1969), Riehl (1969) and Newell *et al.* (1970).

Global patterns of evaporation from the ocean surface

To estimate ocean–atmosphere fluxes of water vapour Jacobs employed the following simple formula, which he developed empirically from foundations laid theoretically by Sverdrup (1937) and Montgomery (1940):

$$E = 0 \cdot 143 \, (e_o - e_z) \, U_z, \qquad \ldots [4.3]$$

where E denotes evaporation (mm day^{-1}), e vapour pressure (mb) and U wind-speed (m sec^{-1}), and the subscripts o and z refer to the sea-surface and height above sea-level z respectively. Thus, using published marine climatic data, Jacobs was able to

(legend within figure:) Isotherms in °C R_N in cal cm^{-2} day^{-1}

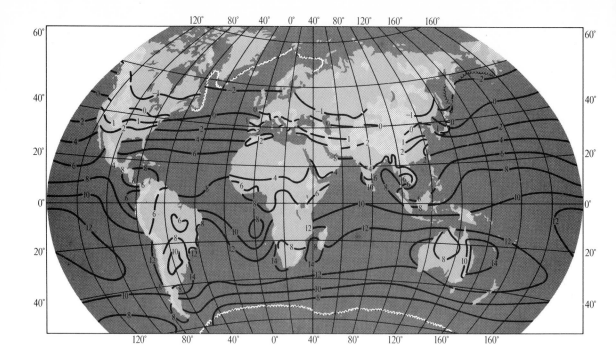

Fig. 4.7. The surface radiation balance in December (units: kcal cm^{-2} month^{-1}).

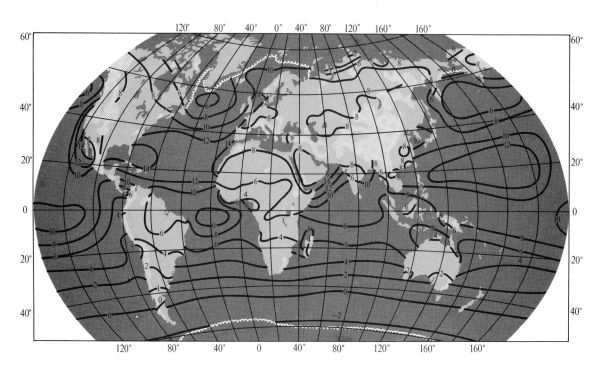

Fig. 4.8. The surface radiation balance in June (units: kcal cm^{-2} month^{-1}).

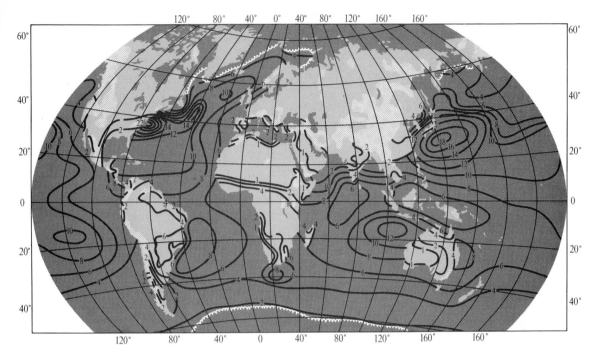

Fig. 4.9. Upward fluxes of latent heat in December (units: kcal cm^{-2} month^{-1}).

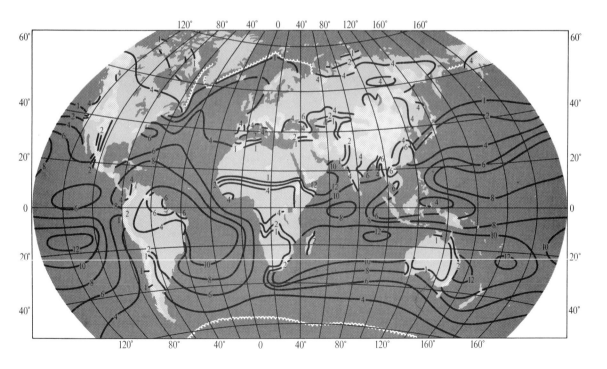

Fig. 4.10. Upward fluxes of latent heat in June (units: kcal cm^{-2} month^{-1}).

prepare charts showing seasonal values of evaporation from 5° squares on the North Atlantic and North Pacific Oceans and thence obtain equivalent values of Q_E by means of the elementary relationship $Q_E = LE$, where L is the latent heat of vaporization of water (assumed by Jacobs to be 585 cal g^{-1}). Budyko adopted the same method of estimating fluxes of latent energy but amended eqn [4.3] slightly to:

$$E = 0\cdot134 \, (e_o - e_z) \, U_z \qquad \qquad \ldots [4.4]$$

Although Budyko (1956) evidently considered the climatic data available in the early 1950s sufficient in quantity and quality to make worthwhile the preparation of charts showing isopleths of E and Q_E (and other components of the energy budget) over all the oceans of the globe, they were in truth still barely adequate for the purpose in most oceanic regions. However, the observational programme of the International Geophysical Year [4] yielded data greatly superior in quantity and quality to any he or Jacobs had at their disposal hitherto, and the maps contained in the revised version of his *Atlas of Heat Balance of the Earth's Surface* (Budyko, 1963) are, accordingly, in the words of Budyko *et al.* (1962), 'much more exact and detailed than the previously compiled maps'. In their preparation Budyko used data for 2,000 stations, including 300 in oceanic areas.

Various general conclusions concerning evaporation from oceans can be drawn from the results of Jacobs and Budyko and their collaborators:

(a) Evaporation rates are almost everywhere greater in winter than in summer (Figs. 4.9 and 4.10). This is due to values of U_a and $e_o - e_z$ tending to be greater during the cold season than during the warm. Budyko (1956) has written:

The increase in heat expenditure for evaporation during the cold season is closely associated with the increasing effect of warm currents during these periods, whereas during the warm season cold currents are most active in this respect, thus lowering the expenditure of heat for evaporation. The actual conditions of heat influx to the evaporation surface of oceans, which are associated with the existence of a powerful horizontal heat transmission in the hydrosphere, represent the main factor in increasing losses of energy for evaporation in the cold period.

(b) The highest seasonal values of Q_E are found near the western boundaries of the North Atlantic and North Pacific Oceans in northern winter. According to Jacobs (1951), values can be as high as $1\cdot14$ g cm^{-2} day^{-1} over the Gulf Stream and $0\cdot94$ g cm^{-2} day^{-1} over the Kuroshio. Although winds over these regions are stronger in winter than in summer, evaporation rates are high in winter chiefly because prevailing winds during that season (Fig. 2.8) continually advect cold, dry, continental air across the warm ocean currents. During summer, in contrast, prevailing winds (Fig. 2.10) blow from southerly points of the compass, so air—sea temperature differences (and hence values of $e_o - e_z$) tend to be small.

(c) Seasonal variations of Q_E are nowhere greater than over the Gulf Stream and Kuroshio Systems. Over eastern parts of oceans seasonal variations are very much smaller, because at all times of the year prevailing winds advect maritime air across the cool ocean currents which flow there. This air has been so thermally and hygrologically modified by a long sea track that in the atmospheric boundary layer over these currents values of $e_o - e_z$ are typically rather small.

(d) Throughout the year there is rapid evaporation in the subtropics and in trade-wind belts. In these latitudes air previously raised to the upper troposphere by cumulonimbus activity in Hadley circulations returns to the lower troposphere to be turbulently moistened and thus provided with the buoyancy which prepares it for ascent once again (see Green *et al.*, 1966). Not only are values of $e_o - e_z$ considerable in these regions but so also are values of U_z, for trade-winds blow steadily with an average strength of Force 3 or 4 (see Crowe, 1950). On the North Atlantic Ocean the mean speed of the trades is about 13 to 15 knots; the strongest trades anywhere are those of the South Indian Ocean, where mean speeds reach 18 knots during winter (i.e. between June and September). In the latter region the evaporation loss of water amounts to about 240 g cm^{-2} yr^{-1}.

Fluxes of sensible heat

To estimate ocean—atmosphere fluxes of sensible heat Jacobs applied a formula first derived by Bowen (1926), expressing it in the form:

$$\frac{Q_H}{Q_E} = 0\cdot65 \, \frac{(T_o - T_z)}{(e_o - e_z)}, \qquad \qquad \ldots [4.5]$$

where T_o and T_z denote temperature at the sea-surface and height z respectively.

Jacobs found the global distributions of values of Q_H and Q_E to be similar, except that 'there exist no tropical areas of (sensible) heat exchange to coincide with the tropical areas of high evaporation'. He also found that Q_H/Q_E (known as *Bowen's ratio*, β) is generally greatest at high latitudes and decreases equatorwards. The average values of β published by Sverdrup (1951) range from about 0·1 in low latitudes to 0·45 at $70°$N and 0·23 at $70°$S. Roll (1965) has explained: 'The difference between the values for the two hemispheres is ascribed to the influence of the large continents of the northern hemisphere from which cold air flows out over the oceans in winter.' During such outbreaks of cold air values of β can become quite large, as (for example) Manabe (1957) found on an occasion when $T_o - T_z$ over the Japan Sea exceeded $10°$C; the associated value of β was 2·3.

Seasonal variations of β are 'substantial', according to Palmén and Newton (1969). As a general rule the ratio is positive in winter and close to zero, or slightly negative, in summer. The variation of β with thermal stability in the atmospheric boundary layer has been discussed by Gordon (1952), whose data Roll (1965) used to prepare Fig. 4.11.

The reliability of Bowen's formula has often been questioned by meteorologists, for a number of reasons. First, the assumption was made in its derivation that the eddy transfer coefficients [5] for fluxes of sensible heat and water vapour are equal, and, second, over large tracts of ocean values of $T_o - T_z$ and $e_o - e_z$ are of the same order of magnitude as errors in measurements of temperature and vapour pressure. Furthermore, evaporation from sea-spray (see Montgomery, 1940) and radiative heat exchange through laminar layers next to the sea-surface (see Sverdrup, 1943) are neglected in eqn [4.5]. Although it is now widely accepted that these two processes are climatically unimportant, there is some evidence that both can be significant on the synoptic scale. Work by Fleagle (1956) indicates that radiative processes are considerable when a very steep temperature gradient is present close to the sea-surface, particularly in relation to the formation of advection fog (discussed later in this chapter), and work by Okuda and Hayami (1959) suggests that evaporation from sea-spray is an appreciable factor in latent-heat fluxes when wind-speeds exceed 15 m sec^{-1} or so.

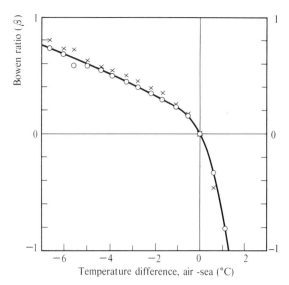

Fig. 4.11. Variation of Bowen's ratio with air–sea temperature difference ($°$C) for Beaufort wind-forces 4 (circles) and 8 (crosses).

Budyko preferred to estimate Q_H directly by application of one of the equations used in the derivation of eqn [4.5], namely:

$$Q_H = \rho_a \, \zeta \, c_p \, (T_o - T_z) \, U_z, \qquad \ldots [4.6]$$

where c_p is the specific heat of air at constant pressure, ζ the drag coefficient (discussed in Chapter 3), and ρ_a the air density. When preparing the maps published in his revised atlas of heat balance components he adopted (Budyko, 1963) the values $\rho_a \zeta = 2·5 \times 10^{-6}$ g cm^{-3} and $c_p = 0·240$ cal $°$C^{-1} g^{-1}, which substituted in eqn [4.6] give:

$$Q_H = 5·18 \, (T_o - T_z) \, U_z \text{ cal cm}^{-2} \text{ day}^{-1}, \quad \ldots [4.7]$$

where U_z is expressed in m sec^{-1}.

It is clear from eqns [4.6] and [4.7] that a large upward flux of sensible heat can be expected in air which is very much cooler than the surface of the sea across which it is flowing, especially when the air-flow is strong. Consequently, it is not surprising to find from the results of Jacobs and Budyko and their collaborators that large seasonal fluxes (see Fig. 4.12) occur over the Gulf Stream and Kuroshio Systems, the Davis Strait, the Sea of Okhotsk and the Barents Sea during northern winter, since at that time of year prevailing winds advect across these waters air chilled over neighbouring land-masses or ice-sheets. Fluxes frequently exceed 300 cal cm^{-2} day^{-1}. As Sellers (1965) has commented: 'The role that these regions play in modifying air masses moving off

81

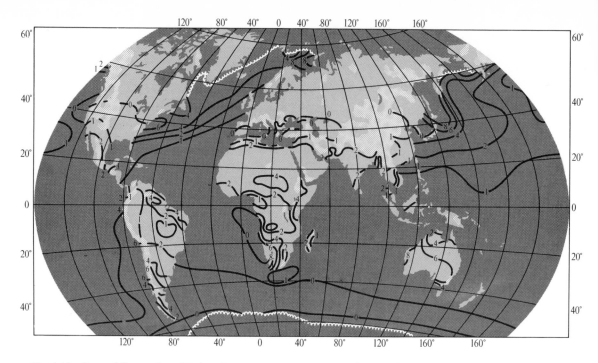

Fig. 4.12. Upward fluxes of sensible heat in December (units: kcal cm^{-2} month^{-1}).

the continents cannot be overestimated. For example, a column of air 1,500 m deep taking up the above amount of heat would be warmed by more than 8°C.'

So great are winter fluxes of sensible heat from north-western parts of the North Atlantic Ocean that values of Q_H exceed 50×10^3 cal cm^{-2} yr^{-1} over the Gulf Stream southward of Newfoundland (in an area bounded roughly by 35° and 45°N, 50° and 70°W), this despite summer fluxes being rather small (Fig. 4.13). Only over tropical deserts are annual fluxes of sensible heat from land surfaces greater (Fig. 4.14).

Lamb (1972) has stated that fluxes from open water near the edge of the ice in latitudes north of 60°N are, averaged over the year, possibly as high as 300 cal cm^{-2} day^{-1}, that is more than 100×10^3 cal cm^{-2} yr^{-1}. Certainly, values of Q_H over the sea can be very high in airstreams originating over ice-sheets. For example, Craddock (1951) found the uptake of sensible heat by Arctic air-masses passing over the sea between Iceland and the British Isles to be typically about 36 cal cm^{-2} hr^{-1} and exceptionally as much as 65 cal cm^{-2} hr^{-1}. To put these figures in perspective he mentioned that the Solar Constant is a little over 115 cal cm^{-2} hr^{-1}. Winston (1955) has reported heating rates of up to 2,210 cal cm^{-2} day^{-1} in northerly outbreaks over the Gulf of Alaska.

According to Budyko (1956), mid-winter 'expenditure of heat by the ocean for turbulent heat emission' to the atmosphere is normally between 4 and 8×10^3 cal cm^{-2} month^{-1} in 'the region affected by the Gulf Stream' and between 2 and 4×10^3 cal cm^{-2} month^{-1} in 'regions affected by the Kuroshio'. Values of Q_H in excess of 2×10^3 cal cm^{-2} month^{-1} also occur over northern parts of the South China Sea and over the Bay of Bengal in December, in association with the development of winter monsoon circulations. In the Southern Hemisphere there are large upward fluxes of sensible heat from water surfaces close to Antarctic ice-sheets, but only over south-eastern parts of the South Indian Ocean and south-western parts of the South Pacific Ocean are annual fluxes large in middle latitudes of that hemisphere. Indeed, Budyko's investigations show annual fluxes of Q_H to be directed downwards (from the atmosphere to the ocean) over some parts of the Southern Ocean. He attributed this occurrence to the frequent advection of warm air over a cold surface. It should be noted, however, that the accuracy of Budyko's evaluations of heat–budget components over the Southern Ocean is rather uncertain, on account of the paucity of observational data from that region at his disposal. Thus, it is probably not surprising to find discrepancies between his analyses and those of

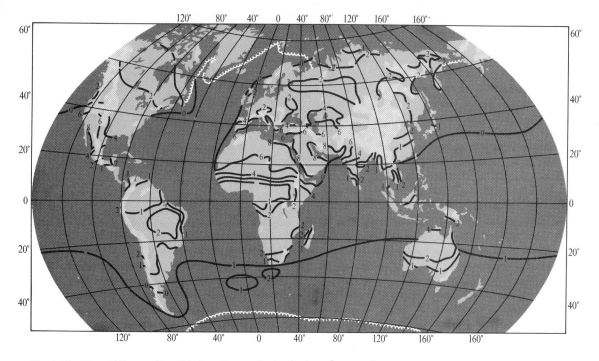

Fig. 4.13. Upward fluxes of sensible heat in June (units: kcal cm^{-2} month^{-1}).

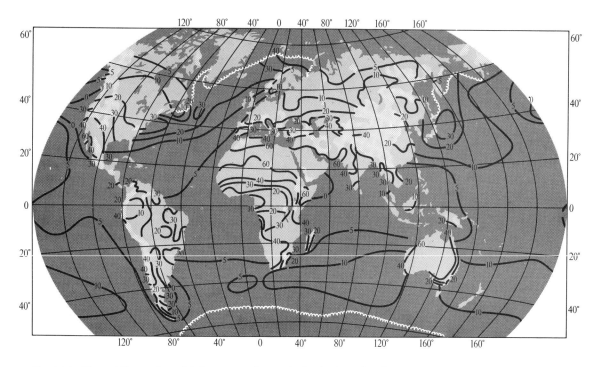

Fig. 4.14. Upward fluxes of sensible heat: annual mean (units: kcal cm^{-2} yr^{-1}).

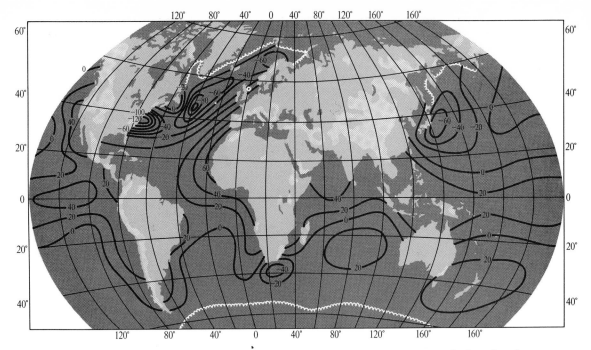

Fig. 4.15. Annual fluxes of heat from the ocean surface to underlying layers of water (units: kcal cm^{-2} yr^{-1}).

other investigators (see, for example: Privett, 1960; Newton, 1972).

During summer months fluxes of sensible heat tend to be small everywhere, except near ice-sheets, and they may even be slightly negative (directed downwards) over cold currents and subtropical oceans. In the tropics and over eastern parts of oceans fluxes tend to be small at all times of year.

Heat fluxes within the ocean

The mean condition of the ocean–atmosphere system varies but little from year to year, so that in climatological calculations of annual energy budgets it is permissible to neglect oceanic storage of heat and so assume $Q_S = 0$. Therefore, once annual values of R_N, Q_E and Q_H are known, corresponding values of Q_{vo} can be obtained from eqn [4.2] as residuals (Fig. 4.15). On time-scales of much more or less than a year, however, fluctuations of Q_S cannot be ignored; secular changes of oceanic heat-storage, although small in magnitude, are probably of some importance in the mechanism of climatic change, and seasonal values of Q_S can exceed $Q_E/3$ (see Newell *et al.*, 1970; Bathen, 1971).

Mindful of the uncertainties in evaluations of R_N, Q_E and Q_H, Malkus (1962) thought it well to

test the residuals carefully for accuracy. This she did by (i) comparing and contrasting Budyko's (1956) estimates of energy-budget components with those published by Jacobs (1951), London (1957) and Sverdrup (1957) and (ii) checking isopleths of Q_{vo} for consistency with the known distribution of ocean currents. She concluded that, 'regardless of details, which should not be pursued too far, the degree of qualitative agreement encourages belief that the residual quantity Q_{vo} has some real physical meaning and exists outside the uncertainties of the calculation'.

It is clear from Fig. 4.15 that large negative values of Q_{vo} are associated with warm ocean currents and large positive values with cold currents. This is not surprising, for over warm currents the sum of Q_E and Q_H annually exceeds R_N greatly, whereas over cold currents values of Q_E and Q_H each tend to be rather small compared with R_N. It must be borne in mind, however, that Fig. 4.15 shows isopleths of oceanic heat-flux *divergence* and not heat *transport* (which can be derived by spatial integration of Q_{vo}). Thus, there is no *a priori* reason why maxima of Q_{vo} should coincide with the main cores of warm and cold ocean currents. Indeed, the largest absolute values of Q_{vo} in the Gulf Stream System (for example) are located westward of the main core of the system. As Malkus has pointed out, the greatest loss of heat by the

84

system probably takes place in its western portions, which are closest to the cold region in the north-west of the North Atlantic Ocean.

The mean annual values of oceanic heat-flux given in Table 4.3 have been obtained by spatial integration of Q_{vo}, under the boundary condition that fluxes vanish in polar regions of the Northern Hemisphere. Perhaps the most notable revelation in this table is that of a small southward flux across the equator.

Unpublished work by Bryan and Webster (1960), discussed by Malkus, indicates that cross-equatorial flow into the Southern Hemisphere is confined to the Pacific Ocean (see also Wyrtki, 1965). They found the cross-equatorial flux in the Indian Ocean to be negligible and agreed with Sverdrup (1957) that in the Atlantic Ocean there is a small northward flux across the equator (see also Bjerknes, 1964, and Emig, 1967).

The annual march of energy—budget components

After examining spatial distributions of components of the energy budget, Budyko (1956) turned his attention to monthly values of R_N, Q_E, Q_H and Q_{vo}, concentrating upon six regions which he considered broadly representative of certain diverse oceanic climates. In summary, his findings were as follows:

1. Near the equator (Fig. 4.16a), R_N and Q_E are of similar magnitude and vary little during the year, while values of Q_H are always small and directed upwards from the ocean surface. Only for a short period during northern autumn is the sum of Q_S and Q_{vo} comparatively large.

2. In contrast to their relative constancy near the equator, energy-fluxes associated with the monsoonal climate of the Arabian Sea (Fig. 4.16b) vary markedly during the year. Values of R_N rise from a winter minimum to reach a maximum in April and May, but thereafter decrease sharply with the onset of the cloudy conditions which accompany the summer monsoon. With the retreat of the monsoon the radiation balance recovers

somewhat before decreasing again to the winter minimum. Rates of evaporation are large in winter because dry north-easterly trade-winds are then present over the Arabian Sea, and they are large in summer on account of the strength of the south-westerly monsoon winds (typically about Force 7 at $15°N$ $70°E$). Since for most of the year there is little difference between T_o and T_z fluxes of sensible heat tend to be rather small. Heat fluxes in the ocean are directed upwards during winter and summer, for during those seasons the sum of Q_E and Q_H exceeds R_N, but during spring and autumn, to quote Budyko (1956), 'great quantities of heat are transmitted from the ocean surface to deeper layers and eventually transported in horizontal directions into other areas of the world's oceans'. The energy budget of the Arabian Sea has also been investigated thoroughly by Colon (1964).

3. At $20°S$ $30°W$ (Fig. 4.16c), a place Budyko believed to be representative of the western periphery of a subtropical anticyclone, R_N changes more or less regularly, reflecting the annual march of R_S; and Q_E and Q_H, which are strongly influenced by the warm Brazil Current, both vary in the opposite sense, reaching maximum values in winter. There is convergence of heat within the ocean in winter months, when $R_N < Q_E + Q_H$, but at other times of year the ocean absorbs heat ($R_N > Q_E + Q_H$). In contrast, at $20°S$ $10°E$ (Fig. 4.16d), on the eastern periphery of the subtropical anticyclone situated over the South Atlantic Ocean, the influence of the cold Benguela Current is such that fluxes of latent heat tend to be much smaller than at $20°S$ $30°W$ and fluxes of sensible heat are directed downwards. There is no month when $R_N < Q_E + Q_H$, so that the ocean gains heat at all times of year.

4. Annual variations in R_N, Q_E and Q_H at two places in middle latitudes are shown in Figs. 4.16e and 4.16f. At both places the surface of the ocean is warm, inasmuch as $55°N$ $20°W$ is located in the decay region of the Gulf Stream System and $45°N$ $160°E$ lies in the path of the Kuroshio, and at both places weather is dominated winter and summer by extra-tropical depressions and

Table 4.3 (after Palmén & Newton, 1969) Mean annual oceanic energy transport across latitude circles. Units: 10^{16} cal min^{-1}. Northward fluxes positive, southward negative.

Lat.	$0°$	$10°$	$20°$	$30°$	$40°$	$50°$	$60°$	$70°$
N	−0·45	1·40	2·17	2·09	1·49	1·08	0·49	0·17
S	−0·45	−2·04	−2·30	−1·94	−1·58	−1·14	−0·45	−0·16

anticyclones. However, the wind regimes at the two places differ considerably; at 55°N 20°W the prevailing wind-direction is westerly at all times of year, whereas at 45°N 160°E there is a marked seasonal wind-shift, from north-westerly in winter to southerly in summer. At both places the annual variation in radiation balance is of large amplitude, with negative values of R_N obtaining in winter, and at both places fluxes of latent heat are large in winter and somewhat smaller in summer. The monsoonal influence is most clearly evident in the values of Q_H. At the Atlantic station fluxes of sensible heat are large in winter and small in summer and directed upwards from the warm ocean surface into the atmosphere at all times of year[6]. At the Pacific station fluxes of sensible heat are also large and directed upwards in winter, due to the frequent advection of cold air of continental origin across the warm sea, but they are directed downwards in summer, for then, owing to the prevalence of southerly winds, the air is usually warmer than the underlying sea-surface. In the words of Malkus (1962): 'It is therefore clear that the value of turbulent heat flux (Q_H) represents an important quantitative index of the influence exerted by monsoonal circulation on heat exchange.' In winter there is convergence of heat within the ocean at both 55°N 20°W and 45°N 160°E (since $R_N < Q_E + Q_H$), and the thermal resources of the Gulf Stream System and the Kuroshio are then drawn upon to a considerable extent. In summer $R_N > Q_E + Q_H$ at both places. At the Atlantic station oceanic convergence amounts to about 50×10^3 cal cm^{-2} yr^{-1}, whereas at the Pacific station $R_N \simeq Q_E + Q_H$ for the year as a whole.

The energy budget of the ocean–atmosphere system

Quantification of the energy budget for the ocean surface and identification of the principal spatial and temporal features of the budget's components are but essential preliminary steps towards complete understanding of the ocean–atmosphere system and, concomitantly, acquisition of satisfactory answers to what Malkus has called 'some vitally important questions to marine scientists'. She wrote: 'Among these are questions concerning the relative importance of poleward heat transport in ocean and atmosphere, the fate of the water-vapour

fuel and its use in driving the air circulations, the creation of wind systems on various scales, and their rôle in exchange and in the maintenance of ocean currents.' To these matters we now turn, examining first the energy budget of the ocean–atmosphere system.

For a column of specified cross-sectional area, extending from the top of the atmosphere into the depths of the ocean, the annual heat budget of the system can be formulated thus:

$$R_Z = L(E - P) + Q_{vo} + Q_{va}, \qquad \ldots [4.8]$$

where P denotes precipitation, Q_{va} the horizontal divergence of heat and potential energy[7] in the atmosphere, and R_Z the radiation balance of the entire column (the difference between short-wave radiation absorbed and net long-wave radiation emitted); the symbols L, E and Q_{vo} retain their aforementioned meanings. Quantitatively insignificant terms can be disregarded, as in eqn [4.2], and moreover, since eqn [4.8] refers to the *annual* heat budget, atmospheric and oceanic storage of heat can also be neglected.

The advent of satellite techniques has made possible the confident evaluation of R_Z, since the data required for its evaluation are nowadays considerably more reliable than formerly — these data relate to factors previously difficult to quantify, namely the absorption of short-wave radiation in the atmosphere, the amounts of various cloud-forms, and the radiative, reflective and absorptive properties of clouds (for discussions of relevant satellite techniques, the following works may profitably be consulted: Vonder Haar & Suomi, 1969; Suomi, 1970; Barrett, 1974). Recent evaluations of R_Z tend to uphold the opinion of Houghton (1954), that the balance is positive equatorward of latitudes 38°N and S and negative elsewhere (see Fig. 4.17). Accordingly, since low latitudes are apparently not steadily warming, nor high latitudes steadily cooling, it can be deduced from eqn [4.8] that oceanic transport of sensible heat and atmospheric transports of sensible heat, latent heat and potential energy must take place. Heat must be exported from zones of positive radiation balance to zones of negative balance. Since the sum of the terms on the right-hand side of eqn [4.8] represents the total heat-flux divergence in the ocean and the atmosphere at a given place, spatial integration of R_Z expressed as a function of latitude yields the total flux of heat across latitude circles. Interestingly, Newell (1974)

Figs. 4.16a–f. The annual march of energy-budget components at (a) 0°N 150°E, (b) 15°N 70°E, (c) 20°S 30°W, (d) 20°S 10°E, (e) 55°N 20°W, (f) 45°N 160°E.

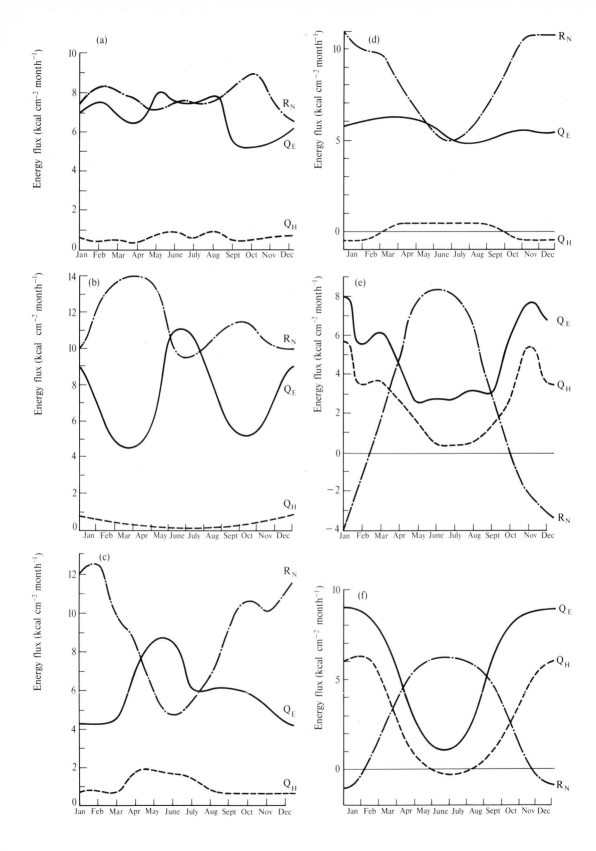

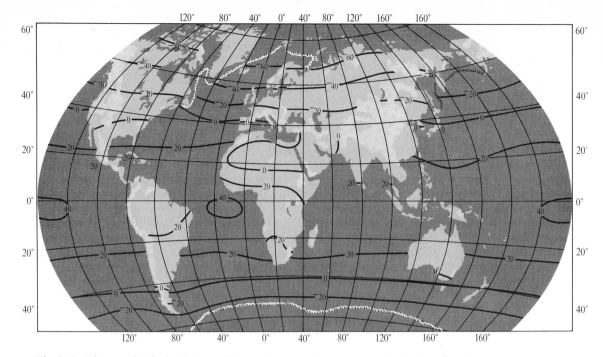

Fig. 4.17. The annual radiation balance of the earth–atmosphere system (units: kcal cm^{-2} yr^{-1}).

has proposed that changes in the proportion of the poleward energy flux undertaken by the atmosphere and the ocean could be a possible cause of Ice Ages.

The contribution made to this flux by atmospheric transport of water vapour can be computed if oceanic evaporation and precipitation distributions are known. Herein lies a problem, however, for it is difficult to measure precipitation accurately from aboard ships at sea (the difficulties have been discussed by WMO, 1962a, and Roll, 1965). Reliance generally has to be placed on extrapolated values of island and coastal precipitation statistics. Not only are such statistics perhaps not representative of adjacent oceanic areas, but also the extrapolation technique is inappropriate for estimating precipitation distributions over oceanic areas far distant from the coasts and islands where precipitation amounts are measured. Furthermore, as Stoddart (1971) found when assessing rainfall data from coral islands in western and central parts of the Indian Ocean, there can be serious doubts over the accuracy of the very measurements upon which extrapolations are based. Local orographic effects may be considerable, he thought, even on low-lying coral islands; many precipitation records are too brief for the long-term average to be reliably established; and it is not uncommon for observers

to be untrained and unsupervised. Work by Barrett (1973) suggests that it may in the future be possible to estimate rainfall amounts from satellite observations of clouds.

Although several decades have elapsed since the publication of Supan's (1898) pioneering attempt to map isohyets[8] over the oceans – an enterprising and creditable endeavour, considering the paucity of stations possessing reliable precipitation records in those days – such is the scope for dispute about oceanic precipitation distributions that aspects of a fundamental nature have yet to be resolved, and there is, consequently, still disagreement over whose isohyetal maps are the most acceptable. For example: Palmén and Newton (1969) adopted the precipitation values given by Meinardus (1934); Rasool and Prabhakara (1965, 1966) made use of precipitation data published by Brooks and Hunt (1930); Neumann and Pierson (1966) chose the data Jacobs (1951) calculated from values published by Wüst (1936); and Malkus (1962) preferred the isohyetal patterns presented by Drozdov and Berlin (1953)!

Discussing the work of Drozdov and Berlin, Malkus mentioned: (i) that they ascertained precipitation amounts by direct extrapolation between stations and did not adjust them for possible orographic influences; (ii) that the isohyetal distributions they drew 'are almost entirely

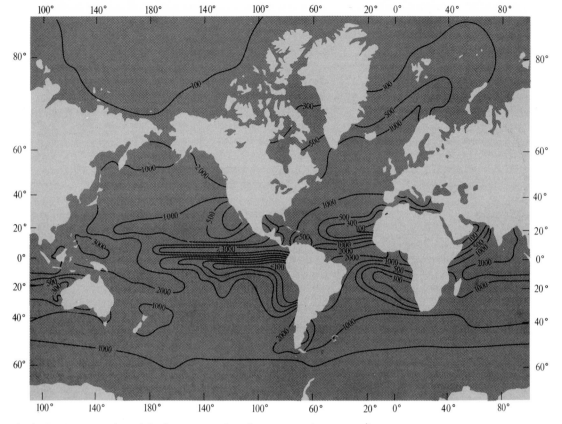

Fig. 4.18. Mean annual precipitation amounts over the oceans (units: mm yr^{-1}).

similar to those of Jacobs, with patterns and maxima which are entirely superposable'; (iii) that energy–budget studies carried out by Riehl and Malkus (1958) encourage belief in the precipitation amounts of Drozdov and Berlin and Meinardus, rather than those of Jacobs, which were reduced 'for presumed coastal enhancement' (the words of Malkus); and (iv) that coastal and island influences may be small, in view of the fact that the majority of precipitation occurs, even in the tropics, in synoptic-scale cyclonic systems. It is thus justifiable to follow Malkus and reproduce in Fig. 4.18 the Drozdov and Berlin chart of the mean annual distribution of precipitation over the oceans.

Knowing R_Z, $L(E - P)$ and Q_{vo} (LE and Q_{vo} are discussed earlier in the present chapter), values of Q_{va} can be derived residually from eqn [4.8]. Thence, by spatial integration, the atmospheric flux of sensible heat and potential energy across parallels of latitude can be computed and the relative importance of the oceans and the atmosphere to the annual heat balance of the globe assessed. Because the values of Q_{va} are so derived,

however, it is, as with the values of Q_{vo} derived residually from eqn [4.2], unwise to draw conclusions from them until their accuracy has been verified. Fortunately, there are sufficient aerological data available for some parts of the world to permit evaluations of Q_{va} to be made directly from observations of atmospheric variables. The results of these evaluations are reassuring, for they indicate that a fair measure of reliance can be placed upon residual values of Q_{va}, despite the uncertainties inherent in estimations of the terms in eqn [4.8]. Indeed, Malkus felt able to assert confidently: 'The average annual global energy transactions are now fairly well-known.' The transactions are summarized in Figs. 4.19 and 4.20.

Inspection of Fig. 4.19 reveals that four zones are distinguishable in each hemisphere:

1. Lying between latitudes 15°N and S (approximately) is the equatorial zone, where R_Z is positive and P exceeds E. Malkus has called this zone 'the atmosphere's firebox', for it contains the most intense energy source in the general circulation of the ocean–atmosphere system. The majority of

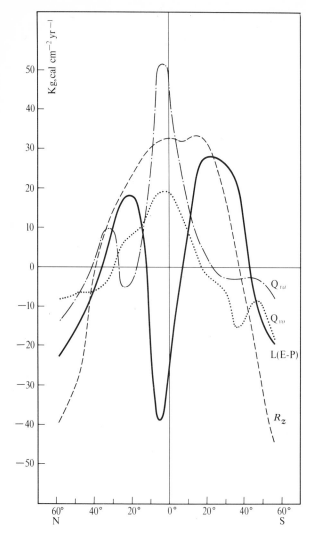

Fig. 4.19. Components of the energy budget as a function of latitutde. For explanation, see text.

the energy stems from the release of latent heat in the cumulonimbus clouds of the Intertropical Convergence Zone and not, as it may have seemed reasonable to suppose hitherto, from the positive radiation balance (see Riehl & Malkus, 1958). It is evident from Fig. 4.19 that the strength of the equatorial energy source is greatest between the equator and latitude 10°N, in accordance with the observed mean annual position of the ITCZ (which is at about latitude 5°N).

2. In the trade-wind zones, located on the equatorward flanks of subtropical anticyclones, values of R_Z are positive (but smaller than in the equatorial zone) and E exceeds P. Values of R_Z and $L(E - P)$ are comparable, and Q_{vo} and Q_{va} are,

accordingly, small in magnitude. Within the trade-wind zones the atmosphere accumulates much of the latent heat which is subsequently released in the equatorial zone.

3. The subtropical ridges of high pressure lie at latitudes 35° to 40°N and S. In these areas the terms of eqn [4.8] are all numerically small, but net poleward fluxes of energy are maximum (Fig. 4.20).

4. Poleward of the subtropical ridges all terms in eqn [4.8] are negative. In middle latitudes, energy is supplied by atmospheric and oceanic advection, and latent heat is released to the atmosphere in precipitation systems. In the Arctic and Antarctic, where the primary energy sinks of the ocean–atmosphere system are located, $L(E - P)$ and Q_{vo} are small in magnitude and Q_{va} is almost as large as R_Z [9].

It is to be expected that further insight into the mechanics of the ocean–atmosphere system will result from an accurate quantification of the seasonal energy budget. To date, however, endeavours to quantify this budget have been so hindered by inadequacy of data that information about seasonal exchanges of energy is still far from complete. Evaluation of oceanic heat storage has proved especially difficult, and such is the lack of aerological data from the Southern Hemisphere that a statement made by Newell *et al.* (1970) remains valid today: 'Atmospheric transports have yet to be derived for the region south of 30°S on a seasonal basis' (see, however, Rasool & Prabhakara, 1966).

Nevertheless, from the results of research into the seasonal energy budget of the Northern Hemisphere certain general conclusions can be drawn. In particular, results indicate that very little energy is transferred across the equator by atmospheric and oceanic motions; the bulk of the radiative energy surplus within a hemisphere is stored in the oceans during spring and summer and liberated into the atmosphere during autumn and winter. According to studies cited by Palmén and Newton (1969) and data published by Newell *et al.* (1970), changes in oceanic storage in the course of a year are greatest near the subtropical ridges (see Fig. 4.21). Palmén and Newton have commented: 'this would be expected from the large seasonal variations of the air masses sweeping over the oceans in these latitudes, as well as from the variations of insolation'.

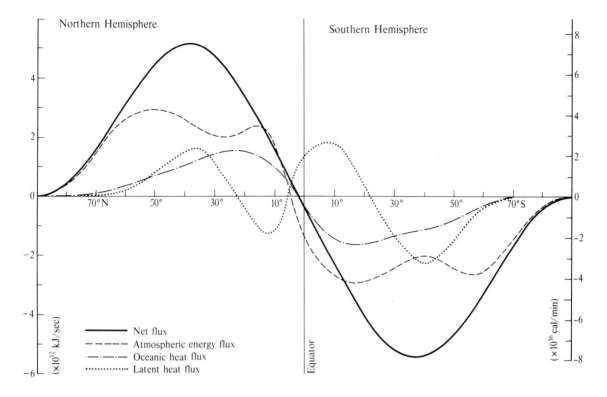

Fig. 4.20. Mean annual fluxes of energy in the ocean–atmosphere system.

Features of the atmospheric boundary layer

From budgetary studies we turn our attention to the microscale physical processes by means of which ocean–atmosphere exchanges of sensible and latent heat are effected in the perpetually turbulent atmospheric boundary layer. As Bunker (1960) has pointed out, the empirical formulae employed by Jacobs, Budyko and others are invaluable for estimating energy-fluxes on a global or hemispheric scale, but they are incapable of providing detailed information about processes operating in the atmosphere immediately above the surface of the sea or about the magnitudes and directions of heat-flows at different heights. To obtain this information, instrumental measurements of turbulent fluxes and adequate supporting theory are required. However, since it is impracticable to make measurements with delicate instruments in the sea conditions accompanying strong winds (see also Chapter 3), present knowledge of turbulent fluxes is largely restricted to that obtained on occasions of light or moderate winds. Roll (1972) thought this 'deplorable, because it is to be expected that the interchange [of energy between ocean and atmosphere] will be

more intense and more important with high wind speeds'.

Although motions in the upper ocean and the lower troposphere are essentially turbulent, there exist at the ocean–atmosphere boundary itself very thin interfacial layers in which laminar motions prevail. The marine layer is about 0·5 mm thick (Neumann & Pierson, 1966) and the atmospheric layer about 1 mm thick (Roll, 1965). Within these layers molecular transfer processes predominate, and vertical gradients of momentum, heat and water vapour are, correspondingly, steep. Separating the laminar layers from the layers in which fully-developed turbulent motions prevail are thin transitional layers in which molecular and turbulent motions are comparable in magnitude.

Of turbulence, Smith (1975) has written:

All turbulence is essentially stochastic, that is it has a basic random content consistent with some bounded probability distribution, whilst at the same time being correlated both in space and in time. Thus the specification of turbulence is largely in terms of statistical means although it is often useful to think of turbulence as made up of eddies of various sizes interacting with the mean

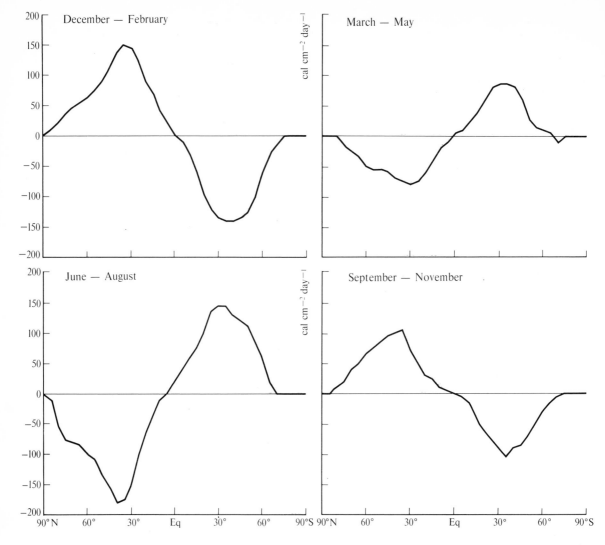

Fig. 4.21. Oceanic heat storage during the periods December to February, March to May, June to August and September to November, according to Newell *et al.* (units: cal cm^{-2} day^{-1}).

flow and with each other. One can associate lines (or tubes) of vorticity with these eddies and in three-dimensional turbulence this eddy interaction results in the mutual stretching of these vortex lines, increasing the local vorticity, reducing the cross-sectional area of the eddies and thereby making them more susceptible to the effects of viscous dissipation leading to ultimate disappearance. . . .

It has proved most convenient and illuminating to describe turbulence in terms of its energy spectrum which displays how the total turbulent energy content is spread out over the whole range from the large energy-input eddies down to the smallest dissipating eddies.

Clearly, in theoretical treatments of so complex a phenomenon mathematical and statistical techniques and manipulations of a very advanced nature are to be expected. Such methods are beyond the scope of this book. Accordingly, we can do no more than draw attention superficially to certain relationships and approaches which have proved useful for ascertaining fluxes of sensible and latent heat in the boundary layer of the marine atmosphere. References to erudite treatises on the subject of atmospheric turbulence are contained in the reviews of Roll (1965), Sheppard (1970) and Smith (1975).

The most comprehensively studied part of the atmospheric boundary layer is that which extends upwards from the laminar layer superjacent to the ocean surface through a height of a few tens of metres and within which it is permissible to assume that turbulent fluxes of heat and moisture are constant with height. In the formulation of these fluxes it is customary for appeal to be made to the *mixing length concept* of turbulence (Prandtl, 1925), according to which, by analogy with the idea of molecular mean free path, discrete 'parcels' of fluid are displaced from their original positions by the action of turbulence and borne through a characteristic distance, known as the mixing length (*l*), before merging again with their surroundings.

Transfer of heat in the atmospheric boundary layer

Suppose that a parcel of air moves upwards through a small distance l, from the level $z-l$, where it originates, to the level z, where it is absorbed into the surrounding fluid. Suppose also that $\bar{q}$ represents the parcel's specific humidity[10] averaged over a short period of time. Since l is small, the vertical gradient of $\bar{q}$ can be assumed constant without prejudicing accuracy unduly. Thus:

$$\bar{q}_{z-l} = \bar{q}_z - 1\frac{\partial \bar{q}}{\partial z}$$

and, because specific humidity is conserved during ascent, $\overline{q'}$, the mean instantaneous deviation of q from $\bar{q}_z$, is given, to a first approximation, by

$$\overline{q'} = \bar{q}_{z-l} - \bar{q}_z = -1\frac{\partial \bar{q}}{\partial z}$$

Thence, by multiplication of $\bar{q}'$ and $\rho_a L \overline{w'}$, where w denotes the vertical velocity (positive upwards), the upward flux of latent heat is obtained:

$$Q_E = \rho_a L \overline{w'q'} = - \rho_a L \overline{w'l}\frac{\partial \bar{q}}{\partial z} \qquad \ldots [4.9]$$

Likewise, and bearing in mind that vertically-moving (unsaturated) eddies change their temperature (T) at the dry-adiabatic rate, it can be shown that the upward flux of sensible heat

$$Q_H = \rho_a C_p \overline{w'T'} = - \rho_a C_p \overline{w'l} (\Gamma_D + \bar{\Gamma}), \quad \ldots [4.10]$$

where Γ_D is the dry-adiabatic lapse-rate (taken as $+ 9 \cdot 8°C \text{ km}^{-1}$) and Γ is the actual gradient of temperature with height. The corresponding formula for the downward flux of momentum is

$$\tau = \rho_a \overline{w'u'} = \rho_a \overline{w'l}\frac{\partial \bar{u}}{\partial z}, \qquad \ldots [4.11]$$

where u indicates horizontal air-speed.

Sellers (1965) thinks it 'very doubtful' that eqns [4.9], [4.10] and [4.11] 'will ever come into general use' for evaluating turbulent fluxes. This may be an over-pessimistic view, but it is certainly difficult to measure the properties of eddies at sea. Not only are there problems of instrumentation, as on land, but also the platform to which instruments are fixed needs to be stabilized. There are two main instrumentation problems: (i) extremely sensitive instruments are required to detect the tiny eddies in air-flows close to the ocean surface, and (ii) prodigious quantities of data accumulate rapidly if the various properties of each and every eddy are measured. The latter problem necessitates the coupling of integrating circuits or computer systems to sensors, so that the estimation of mean fluxes during particular time-intervals may be facilitated.

In practice, it is not necessary to measure the smallest eddies of the spectrum at a given level, because, as Deacon and Webb (1962) put it, 'in the progressive handing down of energy from large eddies to smaller ones, the anisotropy[11] of the large eddies, which enables them to effect momentum or heat transfer, etc., is rather rapidly lost'. Results from observational studies show that, when the atmosphere is unstable or in a condition of near-neutral stability, the highest frequency of eddy passage which need be considered is given, approximately, by $\bar{u}/z$, where z is the height at which measurements of $\bar{u}$ are made. This rule-of-thumb stems from the scale of eddy motion increasing with height.

Correlation of turbulent fluctuations (of an atmospheric property about its mean value at the level of measurement) with vertical components of eddy motion forms the basis of the *eddy-correlation* technique for evaluating turbulent fluxes (see Swinbank, 1951; Priestley, 1959). Alternatively, evaluation may be effected by *profile* techniques. These depend upon the assumption that fluxes are proportional to the vertical gradients of their respective properties and are expressible in the forms:

$$Q_E = - \rho_a L \ K_E \frac{\partial \bar{q}}{\partial z}, \qquad \ldots [4.12]$$

$$Q_H = - \rho_a C_p K_H (\Gamma_D + \bar{\Gamma}), \qquad \ldots [4.13]$$

and

$$\tau = \rho_a K_M \frac{\partial \bar{u}}{\partial z}, \qquad \ldots [4.14]$$

where K_E, K_H and K_M are, respectively, the coefficients of eddy diffusivity for water vapour, eddy conductivity for heat and eddy viscosity. The technical difficulties of applying the profile method at sea have been discussed at length by Roll (1965). The chief metrical problem is due to the vertical gradients of heat and water vapour above the transitional layer being rather small.

Combinations of eqns [4.12] and [4.14] and eqns [4.13] and [4.14] give:

$$\frac{Q_E}{\tau} = -\frac{L\,K_E\,\partial\bar{q}/\partial z}{K_M\,\partial\bar{u}/\partial z} \quad \text{and} \quad \frac{Q_H}{\tau} = -\frac{C_p\,K_H\,(\Gamma_D + \bar{\Gamma})}{K_M\,\partial\bar{u}/\partial z}$$

Introduction of finite differences allows these equations to be rewritten in the forms

$$\frac{Q_E}{\tau} = -\frac{L\,K_E\,(\bar{q}_o - \bar{q}_z)}{K_M\,(\bar{u}_o - \bar{u}_z)}, \quad \frac{Q_H}{\tau} = -\frac{C_p\,K_H\,(\bar{T}_o - \bar{T}_z)}{K_M\,(\bar{u}_o - \bar{u}_z)}$$

Thence, substitution of $\tau = \rho_a\,\zeta\,U_z^2$ (where $U_z = \bar{u}_z$), and $\bar{u}_o = 0$ yields

$$Q_E = \rho_a\,\zeta\,L\,\frac{K_E}{K_M}\,(\bar{q}_o - \bar{q}_z)\,U_z,$$

and

$$Q_H = \rho_a\,\zeta\,C_p\,\frac{K_H}{K_M}\,(\bar{T}_o - \bar{T}_z)\,U_z$$

If, as is commonly assumed, $K_E = K_H = K_M$, it follows that

$$Q_E = \rho_a\,\zeta\,L\,(\bar{q}_o - \bar{q}_z)\,U_z$$

and

$$Q_H = \rho_a\,\zeta\,C_p\,(\bar{T}_o - \bar{T}_z)\,U_z.$$

These are the equations (compare eqns [4.3], [4.4] and [4.6] Jacobs and Budyko found so useful for evaluation of the global energy budget.

For many years the validity of assuming $K_E = K_H = K_M$ has been the subject of disagreement among workers researching boundary-layer fluxes. Much of the disagreement stems from the coefficients not being constant, but varying, in an imperfectly understood manner, with atmospheric stability, altitude above the surface, and aerodynamic roughness of the surface[12]. Moreover, microscale buoyancy forces are involved in fluxes of sensible heat (because density fluctuations are associated with the temperature fluctuations in the boundary layer, air being a compressible fluid), and pressure forces on eddies influence momentum exchange. Thus, as Deacon and Webb (1962) have pointed out, there is no *a priori* reason why K_M should exactly equal the other coefficients. Indeed, the results of investigations carried out by Swinbank (1955) demonstrate clearly that differences between K_H and K_M are sometimes not at all negligible, particularly on occasions of strongly unstable stratification; he found that $K_H > K_M$ in convectively unstable conditions and $K_H < K_M$ in stable conditions. Later work by Charnock (1967) and Swinbank (1968) showed that values of K_H/K_M can exceed 3 in unstable conditions over dry surfaces.

Knowledge of transfer coefficients in maritime air is still rather rudimentary. However, since temperature gradients in constant-flux layers overlying the surface of the sea generally do not differ greatly from Γ_D, except over warm currents in winter and over waters adjacent to ice-sheets at all times of the year, the assumption of $K_H = K_M$ is probably acceptable for most oceanic regions. So far as K_E is concerned, the results of researches cited by Deacon and Webb indicate that equality of K_E and K_M can usually be assumed without serious prejudice to accuracy.

For examining quantitatively the effect of thermal stability on turbulent energy transfer, a non-dimensional parameter known as the *Richardson Number* can be utilized. The parameter (Ri), defined as the ratio of the energy provided by buoyancy forces to the energy produced by shear stress, is commonly given in the form:

$$Ri = \frac{g}{\bar{T}_z}\frac{(\Gamma_D + \bar{\Gamma})}{(\partial\bar{u}/\partial z)^2}, \qquad \dots [4.15]$$

where g is the acceleration due to gravity.

Thus, Ri is negative in lapse conditions, positive in inversion conditions, and zero in conditions of neutral stability. If z is not more than a few metres eqn [4.15] can be rewritten

$$Ri = \frac{g}{\bar{T}_z}\cdot\frac{(\partial\bar{T}/\partial z)}{(\partial\bar{u}/\partial z)^2}$$

and, if finite differences are introduced,

$$Ri \simeq \frac{g}{\bar{T}_z}\frac{\Delta\bar{T}\cdot\Delta z}{(\Delta\bar{u})^2}$$

According to McIntosh and Thom (1969) the equation $K_E = K_H = K_M$ is valid only within the range $-0\cdot03 < Ri < 0\cdot01$.

The formation of sea-ice

So far in the present chapter ocean–atmosphere exchanges of heat have been discussed almost exclusively with reference to air in contact with water. There are in high latitudes, however, extensive areas where ice is interjacent between the waters of the sea and the atmosphere above. Accordingly, when studying the energy budget of these areas it is necessary to take into account heat transfers associated with the formation, growth and decay of sea-ice and the thermal properties of overlying blankets of snow. Passing reference to papers concerned with the energy budget of ice-covered oceans is made elsewhere in the present chapter (p. 107), and the forms and areal extents of sea-ice in the Arctic and Antarctic are described in Chapter 2. Since our attention is presently focused upon microscale processes operating close to the surface of the sea, it is now opportune to consider in some detail the physics of sea-ice. In so doing we reproduce almost verbatim considerable portions of a review article by Walker and Penney (1973).

Water is exceptional in several physical and chemical respects. In particular, the maximum specific gravity of fresh-water occurs at 4°C, whereas freezing begins at 0°C. Consequently, fresh-water becomes stably stratified when its surface is cooled to less than 4°C.

Salinity ($\underline{S}^\circ/_{oo}$) depresses both the temperature at which maximum specific gravity occurs ($T_d\,^\circ$C) and the freezing-point ($T_f\,^\circ$C), according to the approximate empirical relationships:

$$T_d = 4\cdot0 - 0\cdot215\ \underline{S}$$

$$T_f = -0\cdot053\ \underline{S}$$

and it can be shown easily that $T_f = T_d$ when $\underline{S} = 24\cdot7^\circ/_{oo}$.

A typical value of $\underline{S}$ for sea-water is $34^\circ/_{oo}$, and so, because $T_d < T_f$, the upper waters are mixed convectively, by less dense, warmer water rising to replace more dense, cooled water, until freezing occurs. Indeed, in deep sea-water freezing cannot take place unless the surface cooling-rate offsets the effect of vertical exchange of water. Freezing occurs relatively quickly in shallow water because the quantity of water to be cooled is limited by the depth. Moreover, if the salinity is low and is less than $24\cdot7^\circ/_{oo}$, as in river mouths or the Baltic Sea, $T_d > T_f$, the water becomes stably stratified before it freezes and cools more rapidly than convectively-mixed water. Thus, it is evident that ice forms preferentially in water of low salinity and in shallow water.

The foregoing assumes still conditions, but the surface of the sea is normally disturbed by waves, tides and currents, the effects of which may hinder ice-formation considerably and which are difficult to assess quantitatively. Wind affects the rate of evaporation and hence the cooling of surface water (see, for example, Hasse, 1963).

Dissolved salts in sea-water delay the change of state from liquid to solid by interfering with the mechanism of the associated molecular processes, but when solidification does take place these salts are rejected and increase the salinity and density of the remaining surface water. The resulting instability causes the surface water to sink, to be replaced by less dense water from below.

The first visible sign of freezing is the appearance in the water of minute crystals of pure ice. Initially the crystals vary in shape from squarish discoids to hexagonal dendrites (the dendritic form being most common under conditions of rapid cooling), but they soon develop into hexagonal needles, whose widths are about $2\cdot5$ cm and heights $0\cdot1$ to $1\cdot0$ mm.

Following the development of a lattice of ice-crystals on the sea-surface, downward growth of ice proceeds and a cellular structure of ice-crystals forms. The cell-walls are composed of pure water-ice and the cells themselves entrap small amounts of brine. Continued cooling results in a thickening of the cell walls and a concentration of brine within the cells. Ultimately, dissolved material crystallizes out of the brine; for example, sodium sulphate crystals begin to form at $-8\cdot2^\circ$C and sodium chloride crystals at $-23\cdot0^\circ$C (for details of the physical properties of sea-ice see Weeks, 1966, and Weeks & Assur, 1967).

The salt-content of a lump of sea-ice depends upon the rate at which the ice is formed (largely a function of air-temperature and wind-speed) and upon its age. Rapid growth of new ice gives rise to sea-ice of relatively high salinity, but sea-ice which forms slowly tends to be of low salinity. As sea-ice ages, the entrapped brine gravitates out of the ice, which therefore becomes less saline. Old sea-ice tends to be of low salinity but at all times the surface layers of sea-ice tend to be less saline than the deeper layers, as a result of brine drainage [13].

Ice formed in still water is comparatively brittle because the majority of the ice crystals are set vertically. Normally, though, wind and wave action cause ice-crystals to assume a variety of orientations, forming a stronger configuration. Waves and swell further disorder the crystals by fragmenting young ice into small angular pieces (at most a few metres across) which, by collision with each other, become rounded and acquire raised rims. These rounded pieces, known as 'pancakes', adhere to each other to form a continuous sheet.

The growth and decay of sea-ice

Many of the factors which influence the growth of sea-ice are mentioned in this extract from Budyko (1962):

The increase in thickness of the ice cover occurs with the freezing of ocean water at the lower boundary of the ice as a result of cooling. The rate of freezing, under such conditions, is determined by the magnitude of vertical heat flux at the bottom of the ice. The magnitude of the indicated flux may be found by calculating the vertical heat exchange between the lower boundary of the ice and the atmosphere through the thickness of the ice and the snow coat covering it. Such a calculation is based on the solution of the equation of thermal conductivity with the assumption that the temperature of the snow surface is near that of the air, and the temperature at the lower boundary of the ice is $-1\cdot8\,^{\circ}C$. From the solution obtained, it follows that the freezing-rate of ice is determined, basically, by the air temperature and thickness of the ice.

Kolesnikov's (1958) theoretical treatment indicated that the greatest influence upon the growth of ice is the snow-cover, for snow is a good thermal insulator; the rate of ice-growth is dependent upon the depth and compaction of the snow. Wind velocity is apparently much less of an influence, even though it determines the magnitude of turbulent heat exchange in the boundary layer. Kolesnikov pointed out also that 'on clear days ice growth will be more intense than on cloudy days at the same sub-zero air temperature', due to radiation considerations.

Wisely, Kolesnikov was careful to stress the limitations of theoretical studies of sea-ice growth. He recognized that the achievement of strict solutions is essential but drew attention to the need to simplify problems when attempting to formulate them. The characteristic trend of studies aimed at obtaining analytical solutions which are sufficiently simple to be suitable for practical calculations is for there to be a gradual complication of the formulae as more and more factors influencing ice-growth are introduced.

Pounder (1962) has discussed concisely the decay of sea-ice:

The decay of an ice-cover is largely controlled by solar radiation and by the albedo of the surface. The ice stops growing and starts to decay a considerable time before the air temperature rises to the melting point of ice. The later stages of the decay of an annual ice cover in the Arctic are startlingly rapid. It takes place at a season with 24-hour daylight and, until there are significant amounts of open water, under usually cloudless skies. The albedo of the snow cover changes within days or even hours from a value of 0·9, typical of clean snow, to as low as 0·45. The snow cover melts rapidly, leaving wet ice whose albedo is almost as low. An ice cover 8 ft thick can melt completely within 6 weeks.

Perennial pack-ice, he pointed out, breaks into floes in summer and the relatively fresh meltwater drains to the underside of the ice where it freezes, because its freezing-point is higher than that of the saline water.

Advection fog and sea smoke

In their survey of Arctic climate Vowinckel and Orvig (1970) wrote: 'The melting of the pack ice in summer leads to formation of persistent fog and low cloud. More than 100 days per year experience fog at Polar Ocean stations, most frequently in summer and least in winter.' This is a reference to *advection fog*, which forms wherever humid, stable air flows across a surface whose temperature is lower than the dewpoint of that air. According to Vowinckel and Orvig, advection fog is 'particularly prevalent' over Arctic seas from June to September; and in *The Handbook of Aviation Meteorology* (Roberts, 1971) it is stated: 'Over the pack ice and most of the open waters of the Arctic seas, fog every other day during the summer is normal.' Away from the Arctic, widespread advection fog is a frequent occurrence in several regions: over the cold waters of the Labrador Current off Newfoundland (Fig. 4.22); over pack-ice and open waters close to Antarctica; over the Sea of Okhotsk and the Bering Sea; and over the cold waters of the Oya Shio and Aleutian Currents of the northern North Pacific Ocean (for statistical information on the occurrence of fog, atlases of marine climate should be consulted).

Writing about physical processes which may lead to the formation of fog over the sea, Roll (1965) noted that some conceptions widely found in the literature have been formulated imprecisely. He substantiated this point by drawing attention

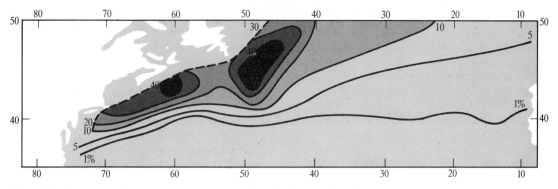

Fig. 4.22. Frequency of fog (per cent) over the North Atlantic Ocean in July.

to shortcomings of the typically-expressed explanation: advection fog is generated because warm, moist air is chilled to its dew-point through contact with a relatively cold sea surface. 'In reality', he wrote, 'an air mass whose dew-point is higher than the temperature of the cold sea surface loses heat as well as water vapour to the sea by the processes of molecular and turbulent conduction and diffusion.' Both heat *and* moisture are transported downwards to the surface of the sea. Accordingly, since the cooling and drying mechanisms are of approximately equal efficiency, both air temperature *and* dew-point decrease, and condensation cannot be achieved unless certain other processes, notably radiative cooling, operate in addition to turbulent exchange (see, especially, Rodhe, 1962).

The thickness of a bank of advection fog is determined largely by wind-strength, given that the surface cooling necessary to produce the fog initially is sufficiently rapid to outweigh the turbulent mixing of fog-laden air with unsaturated air from above. In a wind of moderate strength (5 to 10 m sec^{-1}) it is not unusual for a foggy layer to be several hundred metres thick. When the wind-speed exceeds about 10 or 11 m sec^{-1}, however, stirring of the atmosphere is so vigorous that the cooling influence is overcome and fog dissipates. The temperature lapse-rate in the stirred air then approaches the dry-adiabatic value and the mixing-ratio[14] becomes approximately constant with height. In such circumstances it is usual for condensation to occur in the form of low stratus cloud.

Since the occurrence of widespread advection fog is a function of the disposition of synoptic-scale weather systems, the fog may form or dissipate at any time of day or night and it may persist for many hours, or even a few days. Where, on the other hand, advection fog is a meso-scale

phenomenon there may be considerable diurnal variation of its density. This is particularly so on the coast of California, near San Francisco, where fog results from the drawing of sea-breezes across the cool inshore waters of the California Current.

A different type of marine fog, *sea smoke* (also known as *Arctic sea smoke*), may be observed when the surface of the sea is very much *warmer* than the air in contact with it; 'steam' or 'smoke' appears to rise off the water surface. The phenomenon occurs chiefly during autumn and winter and is most common in Arctic and Antarctic regions and off the eastern coasts of continents, although it has been reported from the Mediterranean, from Hong Kong, from the Gulf of Mexico and even from within the tropics (Bannister, 1948; Starbuck, 1953). Indeed, wherever the necessary temperature contrast between air and sea temperatures develops sea-smoke can form. Any water surface steams if the overlying air is sufficiently cold[15].

The source of cold air is generally close to the water over which the fog forms. For instance: sea-smoke occurs on the micro-scale when very cold air comes in contact with water exposed by ice-breaker action; it occurs on the meso-scale when air cooled over frozen land flows over nearby ice-free waters, such as those of the Icelandic and Norwegian fjords; and it occurs on the synoptic-scale when cold continental air is advected over the coastal waters off eastern Asia and eastern North America. However, it is by no means unknown for sea-smoke to be reported at locations well away from coasts and ice-sheets; that which Hay (1953) observed over the north-eastern North Atlantic Ocean, for example, was formed in air which had travelled nearly 1,000 miles across the ocean.

Saunders (1964) showed, by considering the turbulent transfer of heat and water vapour, that the occurrence of sea-smoke is related to the fact

97

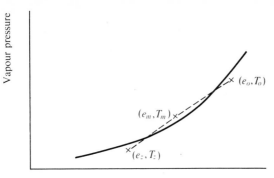

Vapour pressure

$\times\ (e_o, T_o)$

$(e_m, T_m) \times$

$\times\ (e_z, T_z)$

Temperature (T)

Fig. 4.23. The formation of sea-smoke. The curved line represents saturation vapour pressure. For further explanation, see text.

that 'two masses of unsaturated air at different temperatures when mixed together can yield a supersaturated or foggy mixture'. In the case of sea-smoke the two masses concerned are the air close to the surface of the sea and the superjacent cold air.

In the absence of precipitation, external heating and changes of pressure, total amounts of water vapour and sensible heat are conserved during mixing and so, assuming $K_E = K_H$, it is possible to express the mixture temperature (T_m) and vapour pressure (e_m) thus:

$$\frac{e_m}{T_m} = \frac{m_o e_o + m_z e_z}{m_o T_o + m_z T_z}$$

where m denotes mass and the subscripts o and z refer, respectively, to air in contact with the surface and to the overlying cold air (see, for proof: Berry *et al.*, 1945). For all intents and purposes e_o can be regarded as the saturation value of vapour pressure corresponding to T_o, since the molecular diffusivity of water vapour in air at a pressure of about 10^3 mb is sufficiently large ($0 \cdot 23$ cm^2 sec^{-1} at 0°C) to cause air close to the surface of the sea to be nearly saturated.

With reference to the graph of saturation vapour pressure (SVP) against temperature (Fig. 4.23), the co-ordinates (e_m, T_m) lie on a straight line joining (e_o, T_o) to (e_z, T_z). For sea-smoke to appear, (e_m, T_m) must lie above the SVP curve, as shown. It should be noted that the SVP over saline water is some 2 per cent less than over fresh water.

The water droplets which constitute the fog are carried upwards vigorously by convection currents and evaporate into the drier air above. Typically the fog is but a few metres thick and visibility at eye-level and above is not seriously impaired.

Nevertheless, thicknesses of 100 m or so are by no means exceptional and a thickness of as much as 1,500 m is not unknown (for documentation, see Saunders' paper); visibility occasionally falls to as little as 100 m (see, for example, Rubin, 1958). Dense sea-smoke is particularly likely when the ascent of convection currents is limited by a low-level atmospheric temperature inversion.

Cumulus convection

Sea-smoke generally does not form unless $T_o - T_z$ exceeds about 10°C. It is uncommon for this temperature criterion to be satisfied, except in the areas and at the times of year specified above, but it is not at all uncommon for the surface of the sea to be warmer than the overlying air. Lapse conditions prevail over most oceanic regions, especially during winter; and under lapse conditions thermally-generated buoyancy forces collaborate with mechanically-generated turbulent motions to effect vertical exchanges of momentum, sensible heat and water vapour. When turbulent motions more effectively move volumes of air upwards than buoyancy forces it is customary to speak of *forced convection*; when turbulence can be neglected in comparison with buoyancy forces the term *free convection* (or, simply, *convection*) is used.

The height to which atmospheric convection currents extend is determined by the temperature and moisture structure of the troposphere. If, during vertical displacement of a buoyant element (usually known as a *thermal*), condensation occurs, cumuliform cloud ensues. The thicker a convectively unstable layer, the taller associated cumulus clouds grow. Cumulus humilis clouds are of small vertical extent ($H < W$, where H = height of cloud and W = width of its base = 1 or 2 km or so); cumulus mediocris clouds are of moderate vertical extent ($H \simeq W$); cumulus congestus clouds are of great vertical extent ($H > W$); and cumulonimbus clouds make manifest ascent of thermals from the surface of the ocean to the upper troposphere. The majority of cumulonimbus clouds reach, and spread out on encounter with, the thermal barrier of the tropopause and so possess icy anvil-shaped summits. Similarly, the vertical development of cumulus clouds may be arrested by a marked temperature inversion in mid-troposphere, and cloud tops may then, consequently, spread into oval patches. If convection is sufficiently vigorous, as on the eastern flanks of mid-latitude anticyclones, these patches may merge to produce a sheet of stratocumulus cloud.

Over the oceans in temperate latitudes cumuliform clouds are commonly found in anticyclones and in the cool subsiding air to the rear of, and poleward of, depressions. The cumulus clouds are probably formed in thermals, which occur as a consequence of cool air being destabilized when flowing towards warmer regions[16]. In some circumstances convection may be sufficiently strong to produce showers or to produce sheets of stratocumulus clouds.

The previous two paragraphs merely outline the basic features of cumuliform clouds as expressions of free convection over the sea. A detailed treatment of this subject is beyond the scope of this book, and readers who wish to become acquainted with the nature of convective clouds, their varieties of form and their manifold changes of appearance should consult, for example, Scorer and Ludlam (1953), Ludlam and Scorer (1957), Pedgley (1962) and Ludlam (1966b). We confine our attentions to certain aspects of cumuliform clouds in the context of the ocean—atmosphere system.

While on the subject of convection in the marine atmosphere we may mention briefly a phenomenon *associated* with vigorous cumulus and cumulonimbus activity over the sea; we refer to the waterspout. To the eye a waterspout is a funnel-shaped pendant which descends from a cloud base to the sea below. Its diameter may be as little as 2 or 3 m or as much as 300 m, and its height is usually a few hundreds of metres (for details, see Lane, 1968). It travels with its parent cloud at a speed of a few m sec^{-1}, and only rarely does it exist for more than half-an-hour.

In physical terms, a waterspout is a whirling vortex of air rendered visible by condensation of water vapour within it, the condensation being due to the reduction of pressure brought about by centrifugal force in the vortex. The vortex itself is created by vertical stretching of rotating columns of air in the parent cloud (see, for explanation, Scorer, 1958). In and around the base of a spout the sea is agitated and a column, or cascade, of spray is thrown up (see Golden, 1968). According to Gordon (1951), waterspouts occur most frequently in the tropics and subtropics, particularly over the Gulf of Mexico, the Mediterranean Sea and the Bay of Bengal. They are most likely to occur in summer and early autumn (see, for example: Golden, 1973).

The outstanding feature of tropical oceans is the proliferation of *trade-wind cumulus clouds*. Over these oceans, where diurnal temperature changes and air—sea temperature differences are small, cumulus clouds are produced by a mechanism somewhat different from that of the thermals over extra-tropical oceans. This is evident in that the sub-cloud layer in the trades displays an almost dry-adiabatic lapse-rate and a constant mixing-ratio. Such an occurrence is symptomatic of persistent convective mixing (i.e. forced convection) in the steady trade-winds. The bases of the trade-wind cumulus clouds characteristically lie at a level of about 600 m above the ocean surface; and the height to which the clouds grow is governed mainly by the trade-wind inversion, which is normally situated at a level below 3 km. This inversion marks the boundary between the mixed air in the lower troposphere and the subsiding air (of a Hadley-type circulation) in the middle troposphere. A common feature of trade-wind cumulus clouds is their leaning with height, indicating vertical wind-shear. In the cloudy layer wind-speed generally decreases with height.

It has long been recognized that energy conversions in trade-wind belts must be of great importance in respect of the birth and development of tropical cyclonic storms and the maintenance of global-scale atmospheric circulations, but until about three decades ago detailed knowledge of atmospheric conditions over tropical oceans was virtually non-existent and so understanding of tropical cloud and weather phenomena was extremely rudimentary. Since the Second World War this defect has been remedied to some extent, for the structure and behaviour of the tropical atmosphere have been intensively studied, with particular reference to the dynamics of trade-wind cumulus clouds and convective motions in general. Of the earlier studies, those of Riehl *et al.* (1951), Malkus (1952, 1957, 1958, 1962) and Riehl and Malkus (1957) are especially worthy of mention. Details of more recent work have been given by Roll (1965), Garstang (1967), Garstang *et al.* (1970), Augstein *et al.* (1974) and Brümmer *et al.* (1974). Despite the intensive study, however, complete elucidation of processes operating in the moist layer between the trade-wind inversion and the ocean surface has not yet been achieved and endeavours to gain a satisfactory understanding of the tropical atmosphere continue.

Mesoscale organization of cumuliform clouds

A characteristic feature of trade-wind cumulus

clouds is their tendency to become aligned in the direction of the wind, the lines (or rows) frequently extending for several hundreds of kilometres. This phenomenon, revealed by observations from aircraft, is but one manifestation of organization in atmospheric convection.

It has long been known that systematic arrangements of cumuliform clouds can occur — for example, it is quite common to observe a line of cumulus clouds (a *cloud street*) extending downwind of a thermal source, such as a hill-slope facing the sun — but until cloud pictures transmitted from satellites became available the degree and extent of cumulus organization over the ocean was not realized.

Some of the physical conditions necessary for the development of organized convection were identified experimentally many years ago. In the words of Woodcock and Wyman (1947):

Bénard (1900) found that, if a thin layer of liquid, usually free at its upper surface, is heated uniformly at the lower surface, there sets in a regime of polygonal convection cells as soon as a certain critical temperature gradient is reached. The walls of these cells are vertical, and the movement of the liquid is upward in the centre and downward at the periphery. At first, the pattern is somewhat irregular, the polygons formed by a horizontal section through the cells being of different sizes and the number of sides varying from 4 to 7. After a period, however, ranging from one or two seconds to many minutes, depending on the viscosity of the liquid, a much more regular condition is established, in which the polygons approach regular hexagons.

. . . horizontal components of the motion are along radii from the centres of the hexagons, and the downward vertical motion is a maximum along the edges common to three adjacent cells, represented in horizontal section by the vertices of the hexagons. When the liquid is subjected to shear, as a result of horizontal motion, the vertical cells are replaced by horizontal strips, or double rolls, with axes parallel to the direction of shear.

Using smoke as a tracer, Woodcock and Wyman discovered in atmospheric convection mesoscale cellular structures of the kind Bénard generated experimentally in the laboratory, but they concluded erroneously that 'whereas in the liquid cells the upward movement occurs at the centre, in the air cells it occurs at the periphery'. According to Hubert (1966), who has studied thoroughly satellite evidence for mesoscale cellular convection,

i.e. conglomerates of cloud elements organized into cells typically 20 to 100 km in diameter, *two* types of cellular organization occur in the atmosphere. He identified 'open cells', which are characterized by approximately polygonal cloudfree areas surrounded by masses of cloud, and 'closed cells', which are also polygonal, but cloudy, areas, surrounded by strips of clear air. Schematic representations of the two types are shown in Fig. 4.24.

Hubert found that the observed features of mesoscale cellular convection are generally as follows:

(a) Cells form in fields of cumulus or stratocumulus clouds when the atmosphere is heated from the surface beneath. The median cell diameter is about 50 km. The ratio of diameter to depth ranges from approximately 10 : 1 to 100 : 1, the median being about 30 : 1.

(b) Cells are often confined to a shallow surface layer by a strong inversion. Cells may, however, exist in a layer with no inversion, but on such occasions the depth of the convective layer is possibly limited by downward mixing of dry air.

(c) Cells of the open type develop when surface heating is moderate or intense. Weak to moderate surface heating yields closed cells.

Hubert also suggested that cell formation probably requires surface heating to be quasi-steady and vertical wind-shear to be negligible.

It was Hubert's opinion that the atmospheric cells he studied were mesoscale versions of the cells Bénard produced in the laboratory, with the important distinction that molecular transfer processes control the development of cells in laboratory experiments, whereas eddy exchanges regulate the development of atmospheric cells. Hubert came to the conclusion that mesoscale convection is an important aspect of air–sea interaction, 'both because it may represent a scale that makes a significant contribution to eddy exchange processes and because it may be a sensitive indicator of magnitude and distribution of eddy exchange coefficients'.

Organization of convective clouds is a particularly common occurrence in low latitudes (see D. H. Johnson, 1970; Saha, 1971). Mason (1975) has written:

Satellite pictures reveal that these clouds are often formed in giant organized clusters as much as 1,000 km in diameter which last for a few days.

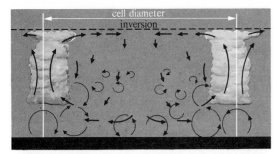

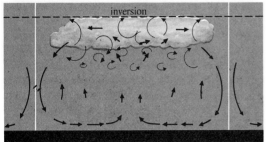

Fig. 4.24. Schematic representation of mesoscale cellular convection. Thin arrows denote turbulent motions and broad arrows general circulatory motions.

They are composed of mesoscale units up to 100 km across which, in turn, contain a number of cumulonimbus cells of 1–10 km diameter, but little is known about their internal structure, organization and dynamics. Discovery of this, and how the formation of the clusters is related to the development of larger-scale disturbances in the equatorial wind field, were the main objectives of a very large, complex international expedition to the tropical eastern Atlantic in June–September 1974.

This expedition is discussed in Chapter 6.

For details of the morphology and thermodynamic structure of cloud clusters, the paper of Martin and Sikdar (1975) should be consulted.

Away from the Intertropical Convergence Zone longitudinal cloud bands are considerably more widespread in the atmosphere than cellular arrangements of cloud. As Kuettner (1959) has pointed out, bands in the lower troposphere are evident as cloud streets in polar outbreaks over mid-latitude oceans and in trade-wind flows; they spiral inwards towards the centres of tropical cyclonic storms and other precipitation systems; and they occur in squall-lines. He found that band structures originate in convective layers, within which wind speeds are greater than usual and wind directions are rather uniform with height. Kuettner

(1971) has written: 'Longitudinal rolls are the preferred convective mode of a flow in which buoyancy forces are counteracted by vorticity forces arising from the vertical shear gradient.' According to Kuettner (1971), longitudinal cloud bands range in length from 20 to 500 km; their spacing varies from 2 to 8 km; the layer containing them is about 0·8 to 2 km in height; and the width to height ratio is between 2 : 1 and 4 : 1. The associated vertical gradient of wind shear is typically 10^{-7} to 10^{-6} cm^{-1} sec^{-1}. Further details of the structure and dynamics of rolls in the lower troposphere have been given by LeMone (1973). A schematic representation of longitudinal rolls, after Burt *et al.* (1974), is shown in Fig. 4.25.

An example of intermediate-scale convective organization

Before proceeding to discuss tropical cyclonic storms, in the development of which mesoscale cellular convection is an element of some importance, we first draw attention to *polar lows*, which are active cyclonic systems embedded in very cold, unstable, polar airstreams. The lows are shallow disturbances, not normally evident in contour patterns above the 700 mb level, and their horizontal extent is small, associated troughs in patterns of sea-level isobars being typically about 200 km across (see Fig. 4.26).

Until recently it was considered that polar lows are purely convective in origin, composed essentially of merged cumulonimbus clouds or cells. It was held that the convective activity results from the heating of cold air-streams over the sea, and it thus seemed reasonable to suppose that formation of the polar lows which pass over or close to the British Isles is in some way due to the heating of very cold northerly airstreams as they flow across the warm waters of the north-eastern North Atlantic Ocean. This hypothesis was apparently supported by the fact that the majority of polar lows occur in winter, the time of year when air–sea temperature differences tend to be greatest.

It is not feasible to test the validity of the convective hypothesis if only routine synoptic and aerological observations are available. The network of observing stations to the west and north of the British Isles is sparse, and many polar lows escape detection until their presence is announced by the arrival of precipitation (usually snow) in Northern Ireland or Scotland. Furthermore, the spacing

101

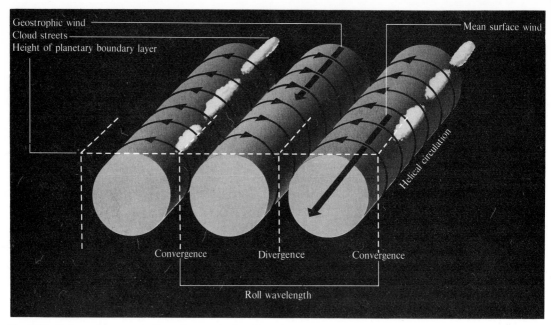

Geostrophic wind
Cloud streets
Height of planetary boundary layer

Mean surface wind

Helical circulation

Convergence   Divergence   Convergence

Roll wavelength

Fig. 4.25. Schematic representation of atmospheric roll vortices.

between radio-sonde stations in and near the British Isles is generally somewhat greater than 200 km. Accordingly, routine upper-air observations provide no more than a modicum of information about circulations in the lows. To procure a detailed knowledge of air motions in the lows, special research methods need to be adopted.

Employing Doppler and conventional radar information, together with routine synoptic data and sequential radiosonde soundings made at the radar station (Pershore, England: 52° 08′N, 02° 02′W), Harrold and Browning (1969) have been able to ascertain the three-dimensional structure of a well-developed polar low which crossed Wales and south-west England during the early hours of 9 December 1967 and to measure associated vertical and horizontal air velocities. They found that most of the precipitation associated with the low formed within a narrow tongue of air ascending uniformly at about 10 cm sec^{-1} and not within cumuliform clouds (see Fig. 4.27). Only over the sea to the west of the low's centre was there vigorous cumulonimbus activity. It was thus apparent that the low was essentially a small baroclinic disturbance, conforming to Ludlam's (1966b) definition of intermediate-scale convection (see Chapter 2).

Harrold and Browning suspected that baroclinic instability in the lower troposphere is a necessary condition for the formation of a polar low. Mansfield (1974) has confirmed theoretically that

this is so and has concluded that the lows develop only when surface winds are light and only where the air flow is not parallel to sea–surface isotherms. Support is thus lent to Lyall's (1972) qualitatively-based deduction that 'the majority of polar lows form within an area to the south and west of Iceland, and usually when the pressure gradient there is slack'. On such occasions the lower troposphere is strongly baroclinic in that area, in association with cold air to the north and north-west and warm air to the south and south-west. It appears, therefore, that polar lows do not develop in the polar airstream in which they are embedded, but move into it later.

Tropical cyclonic storms

Ludlam (1966a) commenced his short but absorbing history of models of the cyclonic storm with the words: 'Amongst the enormous variety of atmospheric phenomena the meteorologist concentrates his studies on the most *energetic*.' This statement is certainly true of tropical cyclonic storms[17], for it is probable that no problem in meteorology has been investigated more intensively in the past few years than that of the formation and development of these storms, the most energetic of all weather systems. The literature is extensive. We confine our attentions to certain aspects of tropical cyclonic storms, particularly

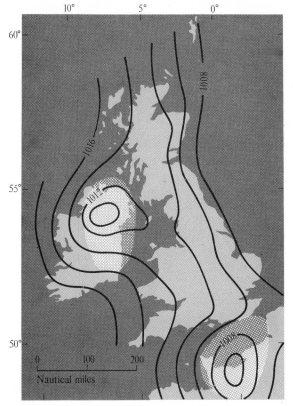

Fig. 4.26. Sea-level isobaric pattern over the British Isles at 1800 GMT on 8 December 1967. Snow was falling within the cross-hatched area.

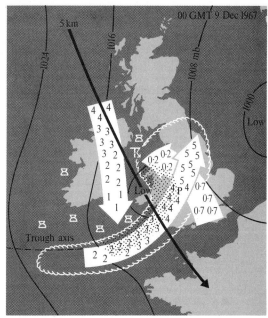

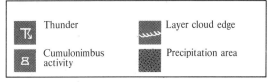

	Thunder		Layer cloud edge
	Cumulonimbus activity		Precipitation area

Fig. 4.27. Composite diagram showing flow relative to the polar low which crossed Wales and south-west England during the early hours of 9 December 1967. Full lines are sea-level isobars, stippling denotes precipitation reaching the ground, and the hatched area delineates the region of extensive layer cloud. Numerals within the flow indicate heights in km.

those concerned with ocean–atmosphere interaction processes. Synoptic and climatological features of the storms are described adequately elsewhere, notably in the publications of Dunn and Miller (1964), Alaka (1968) and Gray (1968, 1975); the meteorology of tropical storms has been reviewed thoroughly by Garstang (1972) and Fendell (1974).

It has been known for several centuries that tropical cyclonic storms develop only over oceanic regions, but not until the middle of the nineteenth century were any of the air–sea interaction processes essential for storm development identified. As Maury put it, in the 1855 and 1858 editions of his celebrated treatise on the physical geography of the sea (in § 943), 'hurricanes prefer to place their feet in warm water'. Although many of Maury's notions about atmospheric systems were untenable and some fantastic, this quotation seems to demonstrate early perception of the now-accepted fact that tropical disturbances do not develop into systems containing winds of hurricane

force (12) unless the surface of the sea is sufficiently warm.

By considering the conditions necessary for the tropical atmosphere to be convectively unstable from the surface of the sea to upper levels of the troposphere, Palmén (1948) was able to show that tropical cyclonic storms can develop only over oceanic regions where the temperature of the sea-surface exceeds about $26°C$ and, hence, that they never form over the South Atlantic and eastern South Pacific Oceans because the surface of the sea is there always too cool (see Fig. 4.28). He pointed out, though, that 'a hurricane, once formed, cannot immediately die' and so may exist over water whose temperature is lower than the critical value.

Tropical cyclonic storms derive most of their energy from release of latent heat of condensation.

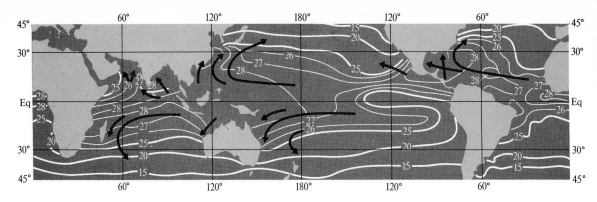

Fig. 4.28. Sea-surface isotherms during the warmest season and paths typically taken by tropical cyclonic storms.

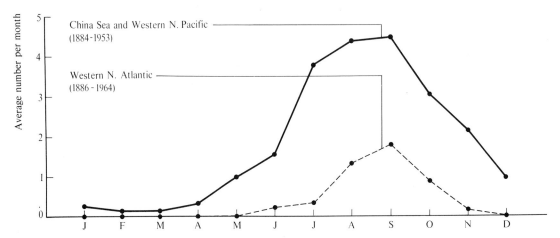

Fig. 4.29. Average number of tropical cyclonic storms per month over the western North Atlantic Ocean (1886–1964) and over the China Sea and western North Pacific Ocean (1884–1953).

Storm activity therefore tends to be greatest over western parts of oceans during late summer and early autumn (Fig. 4.29), since it is in these areas and at this time of year that sea-surface temperatures are highest (see Perlroth, 1969). Cyclones over the Arabian Sea and Bay of Bengal are atypical, however, for they are most frequent in the weeks preceding the onset of monsoon rains in western India and in the weeks following their retreat, that is in May and early June and in November. This inconsistency appears to be unrelated to sea-surface temperature, because extensive areas of the North Indian Ocean are warm enough for tropical cyclogenesis throughout the rainy period in western India. Rather, it is believed by Gray (1967, 1968) that development of tropical cyclonic storms is inhibited by the persistently strong vertical wind-shear which exists over southern Asia and adjacent oceans while the summer monsoon is active (Fig. 4.30).

An apparent *association* between monsoons and tropical cyclonic storms was noted over a century ago by a friend of Maury, one Lieutenant M. H. Jansen of the Dutch Navy, who wrote (see Maury, 1858):

The hurricane season in the North Atlantic Ocean occurs simultaneously with the African monsoon; and in the same season of the year in which the monsoons prevail in the North Indian Ocean, in the China Sea, and upon the Western coast of Central America, all the seas of the northern hemisphere have the hurricane season. On the contrary, the South Indian Ocean has its hurricane season in the opposite season of the year, and when the northwest monsoon prevails in the East Indian Archipelago.

The possibility of there being a relationship between monsoons and tropical cyclonic storms seems to have been disregarded. Certainly it was

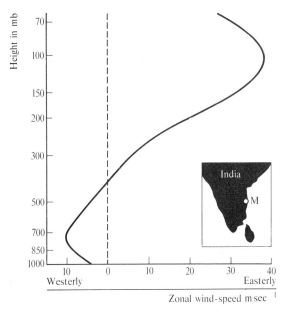

Height in mb (70, 100, 150, 200, 300, 500, 700, 850, 1000)

10   0   10   20   30   40
Westerly   Easterly

Zonal wind-speed m sec⁻¹

Fig. 4.30. Mean zonal-wind profile over Madras for the month of July.

extremely difficult until recently to investigate the idea, on account of the paucity of observations. Perhaps, on the other hand, an association between these two phenomena of the tropical atmosphere was dismissed as fortuitous. However, there is now evidence that it is not so.

Scrutiny of satellite photographs has revealed that many Atlantic hurricanes develop from disturbances which originate over western North Africa or even farther east (Simpson *et al.*, 1968, 1969; Carlson 1969a, b; Garstang, 1972). These disturbances, known as *easterly waves*[18], seem to occur over North Africa only between June and September, the period when an upper-tropospheric easterly flow is established over that area, and indeed over the whole of the tropics from western North Africa to the western North Pacific Ocean (see Flohn, 1964); and this easterly flow is an essential part of monsoon circulations over North Africa and southern Asia (see Flohn, 1964, and Walker, 1972a). The easterly waves themselves have been observed to evolve from atmospheric systems initially no more than mesoscale clusters of cumulonimbus clouds.

Although the *overall* structure and maintenance of tropical cyclonic storms have been understood in general terms for many years, the precise physical circumstances which determine whether or not a tropical disturbance intensifies to become a cyclonic storm are still not adequately known. There are, as yet, many unanswered questions

concerning the highly organized and complex interactions known to exist between air–sea interaction processes, cumulonimbus clouds, mesoscale cloud clusters and synoptic-scale circulations in the tropics (for an impression of progress towards the understanding of these interactions the paper of C. L. Smith *et al.*, 1975, is recommended). It seems from the work of Namias (1973) that atmospheric and oceanic anomalies shaped during antecedent months should also be taken into account.

Not only do tropical cyclonic storms depend for their survival on a supply of sensible and latent heat from the surface of the ocean (see Östlund, 1968; Ooyama, 1969) but also they modify the temperature field of the upper ocean, leaving in their tracks anomalies of sea-surface temperature (see Black & Mallinger, 1973).

Near a storm centre wind-stress causes divergence of surface water and, therefore, a measure of upwelling. Accordingly, the ocean surface becomes anomalously cool along the storm track; in the case of an intense hurricane studied systematically by Leipper (1967) water upwelled from a depth of approximately 60 m within a radius of about 60 km of the storm centre and the sea-surface temperature correspondingly decreased by more than 5°C. Leipper found that convergence of water occurred beyond a radius of about 100 km of the storm centre and that the associated downwelling extended to a depth of 80 to 100 m. Between the regions of upwelling and downwelling he found there was a region within which surface water transported outwards from the storm centre was cooled through mixing with sub-surface water.

The oceanic responses to two vigorous typhoons investigated by Ramage (1972) were essentially in accordance with Leipper's findings (Fig. 4.31). Indeed, the typhoons caused the surface of the sea to be cooled so much that dense, but shallow, advection fog formed over their wakes. This was exceptional, for fog is an extremely rare event in the area concerned, the central South China Sea.

Anomalies of sea-surface temperature induced by tropical cyclonic storms may persist for a few weeks, and since the storms depend upon a supply of heat from the sea it would seem reasonable to expect the movement and intensity of subsequent storms to be influenced by the cool water. However, one of the typhoons Ramage studied proceeded along the track of the other and remained intense until it reached Indochina. Commenting upon this fact, Ramage felt that 'changes in upper-tropospheric ventilation, in

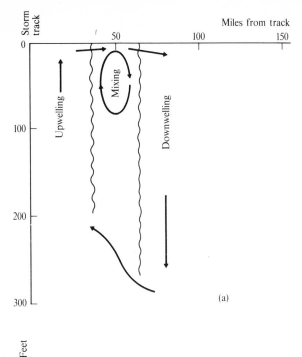

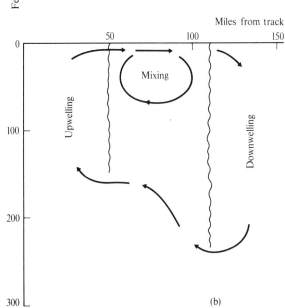

Fig. 4.31. Oceanic responses to the passage of (a) the hurricane studied by Leipper (1967) and (b) two typhoons investigated by Ramage (1972).

subsidence within the eye, and in motion of the atmospheric environment may also be important, at times reducing or neutralizing the effects of surface temperature changes'. On the other hand, the same author (Ramage, 1974), considering the location of near-equatorial troughs of low pressure in relation to sea-surface temperature, has drawn attention to evidence which suggests that development of troughs, and hence of tropical cyclonic storms, is indeed suppressed over water cooled by storm action.

Extra-tropical cyclones

Most tropical cyclonic storms dissipate when they leave the subtropics, but a few become transformed into frontal systems virtually indistinguishable from the depressions which are so common in extra-tropical latitudes. Whereas tropical cyclonic storms produce and maintain their own available potential energy (from liberation of latent heat of condensation), extratropical cyclones are dependent upon the available potential energy of a baroclinic atmosphere. A tropical cyclonic storm accordingly decays when it moves across cool water, unless it comes under the influence of a strongly baroclinic region of the atmosphere, in which case it is regenerated and transformed into an extra-tropical cyclone.

The rôle of ocean–atmosphere fluxes of sensible and latent heat in the development and maintenance of extra-tropical cyclones has been much debated in recent years. According to Petterssen *et al.* (1962), fluxes of sensible heat from the surface of the ocean and liberation of latent heat in clouds 'contribute significantly to cyclone development'. Pyke (1965) agreed with this opinion, and Laevastu (1965) found 'indications that the formation of "heat exchange centres" precedes the birth of cyclones and that the steering of cyclones and the formation of blocking anticyclones might be related to sea–air heat exchange to some extent'. However, the results of Spar's (1965) admittedly crude experiments indicate that diabatic fluxes, including those associated with latent-heat release, are of secondary importance compared with baroclinic effects. Roll (1972) has suggested that ocean–atmosphere fluxes of heat perhaps help to generate baroclinity in the atmosphere and thus contribute to the subsequent cyclogenesis, but he felt that a cyclone, once formed, 'is relatively unaffected by such diabatic fluxes'. Further substantial contributions to the debate have come from D. R. Johnson (1970), Gall and Johnson (1971) and Bullock and Johnson (1972), whose studies tend to support the belief that upward transfer of sensible heat from the ocean surface is an important source of available potential energy in cyclones. As Roll (1970) has commented: 'With regard to the effect the ocean exerts on atmospheric motions of synoptic scale, neither the observational evidence nor theoretical control are satisfactory at present.'

For a thorough review of processes involved in the generation and maintenance of synoptic-scale weather systems in middle latitudes, the monograph of Palmén and Newton (1969) is valuable. Climatic relationships between extratropical depressions and sea-surface temperature are considered in Chapter 5.

Notes

1. The daily radiation totals used in the preparation of Fig. 4.1 are given in the Smithsonian Meteorological Tables.
2. 'Transmission coefficient', also called 'transmissivity', has been defined by McIntosh (1972) as 'the fraction of the radiation intensity incident on a medium which remains in the beam after passing through unit thickness of the medium'.
3. It should be noted, however, that Dr A. F. Bunker, of the Woods Hole Oceanographic Institution, has recently computed surface fluxes of latent, sensible and radiational energy for 500 areas of the North Atlantic Ocean and obtained results which differ somewhat from those of Budyko. Details of fluxes over western parts of the North Atlantic Ocean are contained in a Woods Hole Technical Report (Bunker, 1975). Particulars of fluxes over the entire North Atlantic Ocean were given by Bunker at an international conference on Ocean–Atmosphere Interactions, which took place in Seattle between 30 March and 2 April 1976 (see *Bull. Amer. Met. Soc.*, **57** (1976), 121–55).
4. The International Geophysical Year (IGY) was an ambitious research enterprise involving international cooperation in a world-wide programme of observation in nearly all branches of geophysics. Although the IGY formally extended from 1 July 1957 to 31 December 1958 some sections of the observational programme commenced well before July 1957 and some continued through 1959. The aims and aspirations of the IGY have been discussed in detail in a collection of essays edited by Bates (1964).
5. These coefficients refer to the transfers of heat and water vapour effected by turbulent eddies. The coefficients are: eddy conductivity of heat (K_H) and eddy diffusivity of water vapour (K_E). Their units are $[L^2 T^{-1}]$.
6. Robinson (1966) has discussed the surface energy balance at another Atlantic station, $62°N\ 33°W$.
7. It is necessary to include transport of potential energy because air is compressible, and energy conversions between sensible heat and potential energy can therefore occur.
8. Isohyets are lines of equal precipitation amount.

9. The energy budget of the Arctic Ocean has been investigated thoroughly by Fletcher (1965), Badgley (1966), Vowinckel and Orvig (1966, 1970) and Budyko (1974). Little is known of the energy budget over the Southern Ocean and Antarctica, because of temporal and spatial insufficiency of data.
10. Specific humidity is defined as the ratio of the mass (m_v) of water vapour to the mass ($m_v + m_a$) of *moist* air in which the vapour is contained, i.e.

$$q = \frac{m_v}{m_v + m_a}.$$

11. Isotropic turbulence is determined solely by the nature of the surface over which the wind blows and develops when the atmosphere is in a thermal state of neutral stability. Anisotropic turbulent motions exist when the thermal stratification of the atmosphere is stable or unstable and buoyancy forces suppress or enhance turbulent exchange, as the case may be.
12. According to McIntosh and Thom (1969), an aerodynamically rough surface is one on which 'the individual roughness elements, such as blades of grass, penetrate far enough into the region of turbulent flow to ensure that the shearing stress on the surface is made up of aerodynamic drag forces acting on each one of them individually, rather than on an all-enveloping laminar sub-layer'.
13. The specific heat of saline ice over the temperature range $-23°C$ to $0°C$ has been investigated empirically by Dixit and Pounder (1975).
14. Mixing-ratio is defined as the ratio of the mass of water vapour to the mass of dry air with which the vapour is associated.
15. The term 'sea-smoke' is used when the fog occurs over saline water; 'steam-fog' refers to exactly the same phenomenon over fresh water.
16. The remarkable fact that groups of seagulls can *intentionally* initiate thermals in the marine atmosphere has recently been reported by Woodcock (1975).
17. Otherwise known as 'tropical cyclones' or 'tropical revolving storms'. Regional names are 'hurricane', 'cyclone' and 'typhoon'. Some authors use the term 'hurricane', wherever the storms occur.
18. An easterly wave is a trough of low pressure extending from the equatorial belt of low pressure into the trade-wind flow.

Chapter 5

Thermal behaviour of the ocean–atmosphere system and climatic responses

It is a central theme of this book that the atmosphere and the oceans form a coupled energy-system, and we emphasize that physical processes taking place at the boundary between them are of great importance. In particular, changes of temperature at the sea-surface have profound meteorological and climatological consequences. These consequences are discussed in this chapter.

Monitoring and forecasting ocean temperature

The temperature of the sea-surface is typically about $0 \cdot 5°$C lower than the temperature immediately below the surface, this lower 'skin' temperature being caused by evaporative cooling at the sea-surface. Sea-surface temperature (SST) depends partly on physical processes occurring within the sea itself, such as convective stirring and turbulent mixing, and partly on the influence of the overlying atmosphere. Until quite recently it was assumed that the SST was rather homogeneous over wide areas, with slow and steady horizontal variations, but increasing availability of data has revealed complex patterns, with warm and cold areas in close juxtaposition, especially in regions where divergence and convergence of currents is taking place. These regions of strong horizontal temperature gradients are called *oceanic fronts*.

SST is normally measured from vessels, but there are various sources of error in data so derived (see Ball, 1954; WMO, 1954; Saur, 1963), and increasing use is nowadays being made of infra-red thermometers, or pyrometers, carried in aircraft or satellites, to obtain a synoptic picture of SST over large areas. These devices have an accuracy at present of about $\pm 2°$C and depend upon the conversion of long-wave radiation from the sea-surface into a measure of temperature (Smith *et al.*, 1970).

Persons concerned with naval matters and fisheries, as well as meteorologists, require forecasts of SST behaviour, and the US Navy Fleet Numerical Weather Facility in Monterey, California, has been among the pioneers of such forecasts. About 1,200 SST reports are available every 12 hours from the Northern Hemisphere, but this data density is too low for reliable systematic analysis and it has been found necessary to use data for 3½ days in a forecast. This is satisfactory, provided that the analysis scheme allows some indirect weighting of the data according to age. To prepare hemispheric charts, averaging of sea-temperature over one- and five-degree squares is undertaken, but, if the data are sparse or unevenly distributed in space and time, different subjective weighting techniques can be employed. Because the sea-surface reacts relatively rapidly to changes in the overlying atmosphere the oceanographic forecaster makes use of the daily surface weather analysis as well as the observed sea-temperature data. SST forecasting at Fleet Numerical Weather Facility is based on physical cause and effect principles. In areas of particular interest, such as coastal areas or current boundary regions, a small-scale (zoom) analysis is carried out if data density allows. A zoom analysis with a grid mesh scale of 20 nautical miles can depict small-scale features of the SST distribution.

The vertical temperature structure in the ocean to a depth of about 450 m is normally investigated by the use of the expendable bathythermograph (BT). This is a bomb-shaped device containing a thermistor at its nose and a spool of thin copper wire attached to a deck unit which records variations in temperature as a function of depth. The BT falls at a constant rate. A simpler, older instrument is the reversing thermometer, which turns upside down at the point of observation, breaking the mercury at a constriction in the tube. A good review of these and other instruments, together with observational techniques, can be found in Roll's (1965) book. Since 1962 twice-daily bathythermograph observations have become part of the normal routine aboard British Ocean

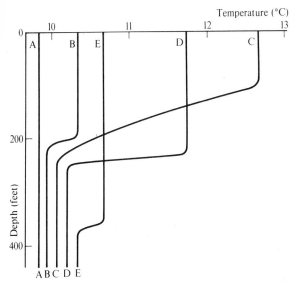

Fig. 5.1. Typical vertical temperature distribution in the North-East Atlantic (a) end of winter, (b) spring, (c) summer, (d) autumn, (e) early winter.

Weather Ships and research vessels, and some fishery and merchant vessels also take these observations. There is no international system in operation for the collection and exchange of BT records as yet.

The broad pattern of the thermal structure of the eastern North Atlantic Ocean has been described by J. D. Perry (1968). The normal vertical temperature distribution consists of an isothermal mixed layer beneath which is a layer in which temperature decreases rapidly with depth, that is a *thermocline*. Seasonal variations occur in the typical vertical temperature distribution as can be seen in Fig. 5.1. At the end of winter the sea is typically isothermal to a depth of more than 150 m, as a result of convective stirring caused by winter storms, but in summer accumulation of heat in the upper ocean can result in a thermocline at a depth of 30–60 m. In early summer the thermocline may be destroyed by a gale, but later in the summer the vertical temperature gradient is such that even the strongest winds cannot provide the increased mechanical effort needed to overcome the density discontinuity. Once this stage is reached the thermocline intensifies rapidly, since incoming heat is distributed within the limited layer above the seasonal thermocline. Perry noted that 'the seasonal cycle is typical of temperate latitudes and is superimposed on the more pronounced oceanographic features such as the Gulf Stream'.

The depth of the mixed layer and indeed the detailed thermal structure varies from year to year, depending upon prevailing meteorological and oceanographic conditions. The seasonal thermocline, in particular, varies widely in duration and intensity from year to year. Temperatures fall within the thermocline to low values, and at about 2,000 m or more the water is close to $0°C$ in all oceans.

Synoptic oceanography for fishery purposes requires that forecasts be made of the oceanic thermal structure. Laevastu and Hela (1970) noted that, while some pelagic fish are found above the thermocline, others frequent the layers of the thermocline itself. Since temperature affects the passage of sound markedly, for naval, and especially for submarine operations, it is desirable to predict the temperature structure. The US Navy has developed ASWEPS (Anti-Submarine Warfare Environmental Prediction System) for this purpose.

Annual, seasonal and monthly-mean data, especially of SST, are readily available in atlas form (an excellent bibliography has been assembled by Wolff & Cartensen, 1965). Because different amounts of data have been used, different numbers of years included in the averages, and different methods of computation employed to prepare the presentations, comparability from one atlas to another is often difficult, and evidence points to considerable differences between the values of monthly-means prepared by different workers. The World Meteorological Organization (WMO, 1962b) has published a set of climatological normals. A few atlases show the mean spatial variations of temperature at given depths (for example: US Navy, 1966; Schroeder, 1963).

Charts of mean SST (Fig. 5.2) show that the expected variations of SST with latitude are much modified by ocean currents. Sea-temperatures in the Southern Hemisphere are generally somewhat lower than those in the Northern Hemisphere because of differences in the character of the prevailing winds and the effects of the large ice-covered Antarctic continent. The highest sea-surface temperatures on record appear to have been recorded in the Persian Gulf, where $34°C$ has been noted. Minimum SST may be defined as the freezing point of sea water, which is a function of salinity (see Chapter 4). The average position of the maximum oceanic surface temperature, the oceanographic thermal equator, is located at about $5–10°N$.

The annual range of SST over most of the oceans of the globe is less than $5°C$, but annual

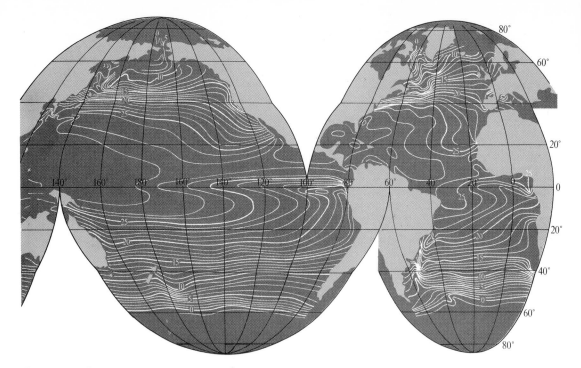

Fig. 5.2. Surface temperature of the oceans (°C) in July.

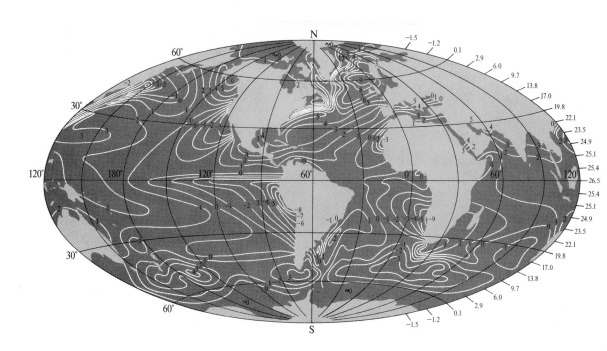

Fig. 5.3. Annual mean sea surface temperature anomalies (°C), expressed as deviations from the latitudinal average.

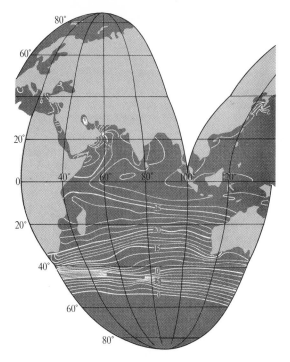

Long-period sea-temperature changes have been observed in many oceans. Since, however, there exists a fairly complete record of North Atlantic Ocean temperatures since 1876 we now consider the secular variations that have occurred in that ocean.

North Atlantic sea temperature record

The scattered records of Danish and Icelandic commercial vessels were collected by the Danish Meteorological Institute and published by Ryder (1917) in the form of monthly means for one-degree squares for the period 1876–1915. Smed (1947–60 and 1962–5) has continued this work and analysed sea-temperature variations in fourteen areas of the North Atlantic (see Fig. 5.5). The quality of these observations of water-temperature does not meet rigorous standards. Highly dissimilar oceanic conditions are found within some of the areas and the statistical weight of the average values was not the same for each year because the number of actual observations varied. The observations were also not uniformly distributed over the water area.

A decrease of temperature occurred in most areas in the first few years of the period, after which temperatures rose somewhat, before again falling and reaching a minimum about 1920. A rapid increase then took place, with temperatures reaching a maximum in the early 1930s in the western Atlantic and in the 1940s and 1950s in the eastern Atlantic, since when a further lowering has taken place. This most recent fall of sea temperature can most easily be monitored by reference to the record of the Ocean Weather Ships (OWS), the locations of which are shown in Fig. 1.6 in relation to the major ocean currents. At least three major variants can be found in the trend of sea temperature since 1951:

1. At OWS D and E a rapid fall of temperature during the early and mid 1950s was followed by a period of rather more stable temperatures, but still with a downward trend.

2. In the eastern North Atlantic, at OWS I, J, K and M, temperatures have fallen markedly since the end of the 1950s, with the lowest temperatures occurring during the 1970s.

3. At the more north-western vessels, OWS A and B, sea temperature actually increased until the mid 1960s, since when a sharp fall has been observed.

ranges are higher in land-locked seas and in continental shelf waters, approaching 20°C near Korea. The sluggish response of the temperature of the sea-surface to heat gain and loss is reflected in the fact that maximum and minimum temperatures occur, respectively, in August–September and February–March, rather than immediately after the solstices, as in the case of air temperature in the continental interiors (Prescott & Collins, 1951). By portraying SST as deviations from the average of each geographical latitude (using the same techniques as for producing isonomalous air temperature maps), anomalous warmth and coldness is brought into focus. The map of Dietrich and Kalle (1957) (Fig. 5.3) shows that the waters north of the British Isles are more than 9°C warmer than is normal for the latitude, the largest positive anomaly in any ocean. The contrast between the relatively warm eastern side and the cold western side of the high latitude oceans can also be clearly seen. Global SST charts are not representative of coastal waters, straits and semi-enclosed sea areas. Occasional investigations in such areas have been carried out, Lumb (1961), for example, finding that around the British Isles the magnitude and horizontal extent of the coastal gradients of sea temperature can be considerable. An example of the complex patterns that may occur in summer is shown in Fig. 5.4.

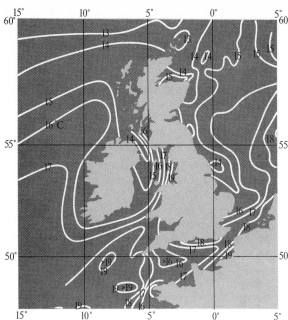

Fig. 5.4. Sea surface isotherms for August 1955.

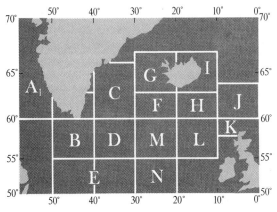

Fig. 5.5. Location of fourteen areas designated by Smed in sea-surface temperature analysis.

It seems that the amplitude of average annual water-temperature variations between the warmest and coldest years varies in different parts of the North Atlantic, being greatest in south-western parts of the North Atlantic and least in the Norwegian Sea. Standard deviations of monthly values of water temperature are 10—20 per cent of the mean annual range in most parts of the ocean. Advection of cold water and of airmasses of polar origin can occur more readily in western parts of the ocean. The decrease in sea temperature in recent years has particularly affected the warm-water months of July—September; for example, at

OWS 'D' the mean August value declined by 2·4°C between the periods 1951—5 and 1966—70, possibly because persistent cyclonicity has led to upwelling of cold water, with increased cloudiness and windiness leading to a reduction in radiation receipt. This is a surprisingly large change over such a short time-span, and it has been suggested (Wahl & Bryson, 1975) that it represents a fall in temperature which might be equivalent to one-sixth of the difference between Ice Age sea temperatures and those of the present, and that values of surface temperature in the North Atlantic today are similar to those during 'the Little Ice Age'.

During the period of declining sea temperatures the mean annual atmospheric circulation anomalies have shown above-normal pressure in the zone 55°—70°N and south-westward displacement of a negative pressure anomaly over the North Atlantic Ocean. Since the zone of greatest temperature decrease lies along the northern edge of the Gulf Stream drift, Wahl and Bryson have suggested that there has been a south-eastward shift of this current and a corresponding advance of colder Labrador Current waters in the Grand Banks area. Anomalous northerly and north-westerly winds over the western North Atlantic could well be responsible for this displacement.

Persistence of sea temperature anomalies

Once anomalies of sea temperature exist they have a strong tendency to persist. The ocean is a much more conservative thermal body than the atmosphere, largely as a result of its high specific heat. The most frequent duration of an anomaly in the North Atlantic is 7—12 months, but at OWS J spells of over 50 consecutive months with positive anomalies have been noted. There are two peak periods of high inter-monthly correlation — the winter and summer months — while the intervening spring and autumn months show lower levels of correlation. The months of peak correlation coincide with the two seasons of most stable vertical temperature distribution (J. D. Perry, 1968). Sea temperature anomalies formed during the two seasons of high inter-monthly correlations are often destroyed during spring and autumn, but anomalies formed at the commencement of winter or summer are likely to persist through to the following spring or autumn. It is interesting to note that Craddock and Ward (1962), investigating air-temperature anomaly relationships in Europe and

112

Western Siberia, found that the pattern of association was strongest in the periods December–April and May–September. It is possible that large-scale atmospheric controls lead to high levels of persistence in both oceanic and atmospheric temperatures at certain times of the year, or there may be some cause-and-effect relationship between the establishment of persistence in one media and its initiation in the other. Namias (1972) has noted that resurgence of anomalies from one cold season to another can occur because the sea has the ability to store warm or cold water masses generated in a deep mixed layer, by covering them with a shallow layer of unrepresentative water during summer, and then resurrect them by renewed stirring during the next stormy winter season. The orderly movement of warm and cold anomalies has been discovered in the North Pacific Ocean (Favorite & McClain, 1973), the anomalies moving around the North Pacific gyre.

Short-period variations of sea temperature

Advection by currents, heat exchanges between sea and atmosphere, mixing by wave action, and convective stirring, can all lead to day-to-day changes of sea temperature, which in many areas of the ocean can amount to as much as $1 \cdot 5^\circ$C over a 2-day period. If several factors act together to change temperature in the same sense even more rapid changes can occur, and in some areas the magnitude of these short-period changes of temperature can equal the total annual range. In general, water temperature decreases with increasing wind speed, the amount varying seasonally but being largest in winter and spring and probably being caused by increased turbulent mixing of the water body. In Table 5.1 it can be seen that low latitudes are more susceptible to large diurnal changes than high latitudes, especially when winds are light and skies clear. On the western side of the ocean the interdiurnal change is $0 \cdot 5^\circ - 1 \cdot 0^\circ$C, in the central part about $0 \cdot 5^\circ$C and on the eastern side smaller still. Manier and Moller (1961) have shown that at OWS A, I, J, K and M maximum water temperature occurs when winds are between south and east. Since water temperature decreases with increasing wind speed, the strongest diminution occurs in spring, being caused by increased dynamical mixing at higher wind speeds.

Table 5.1 Average diurnal variation of sea-surface temperature ($^\circ$C) in summer and winter at various latitudes in offshore areas; after Laevastu and Hela, 1970.

Latitude zone	Wind force Beaufort 0, 1 and 2		Wind force Beaufort > 6	
	Clear	Cloudy	Clear	Cloudy
Summer (April–September)				
$0^\circ - 20^\circ$	1·5	0·8	0·2	0·1
$20^\circ - 40^\circ$	1·0	0·3	0·1	0
$40^\circ - 60^\circ$	0·5	0·1	0	0
$>60^\circ$	0·2	0	0	0
Winter (October–March)				
$0^\circ - 20^\circ$	1·2	0·5	0·1	0
$20^\circ - 40^\circ$	0·5	0·1	0	0
$40^\circ - 60^\circ$	0·1	0	0	0
$>60^\circ$	0	0	0	0

Temperature changes in the sea often follow changes in the air, although the magnitude of the change is much less. In autumn the fall of sea temperature is largely independent of air temperature changes, as mixing takes place between the surface waters and cooler layers below.

Classification of sea temperature anomaly fields

As a result of their search for relationships between oceanic temperature anomalies and atmospheric circulations, meteorologists have recognized the desirability of classifying sea temperature anomaly patterns. Ratcliffe's (1971) North Atlantic classification pays particular attention to the sign of the anomaly of temperature in the area between $35^\circ - 50^\circ$N and $40^\circ - 60^\circ$W and recognizes a total of ten types. The full classification is given in Table 5.2.

For some months several classifications are possible, and, as a first step in developing a more objective classification, attempts have been made to investigate the statistical structure of anomaly fields using principal-component analysis. Figure 5.6 shows the first three eigenvectors derived by Vacnadze *et al.* (1970) for the winter and summer seasons. The analysis suggests that there is a different disposition in the importance of particular fields between the seasons. In a separate, but similar piece of work, Vladimirov and Nikolaev (1970) found that the variety of possible fields can be described by five classes and these are closely related to circulation processes in the atmosphere.

113

Table 5.2 Classification of North Atlantic sea temperature anomaly patterns.

Type	Description
CP5	Cold anomaly exceeding 1°C centred near 50°W
CPE	Cold anomaly exceeding 1°C between 40° and 50°W
CPW	Cold anomaly exceeding 1°C between 50° and 60°W
WP5	Warm anomaly exceeding 1°C centred near 50°W
WPE	Warm anomaly exceeding 1°C between 40° and 50°W
WPW	Warm anomaly exceeding 1°C between 50° and 60°W
DZ	Warm anomaly north of 45°N Cold anomaly south of 45°N
EZ	Cold anomaly north of 45°N Warm anomaly south of 45°N
MWW	Ocean warmer than usual west of 30°W
MCW	Ocean colder than usual west of 30°W

Air–sea temperature difference

The equation by means of which fluxes of sensible heat are computed contains a term which refers to the air–sea temperature difference (see Chapter 4), so the size and spatial variation of this difference is an extremely important factor. Variations of temperature in the sea tend to have a greater effect on fluxes of sensible heat in summer than in winter, when large meridional temperature gradients and vigorous horizontal exchanges of air-masses are common. Rodewald (1972) has noted that there have been systematic differences in the distribution of changes in the air–sea temperature difference between 1951 and 1960 and 1961 to 1970 in the North Atlantic:

(a) In the north, at OWS M, the mean difference increased from 1·9°C to 2·2°C, probably due to increased advection of cold air southward over the Norwegian Sea.

(b) At OWS D the situation was the reverse of that at OWS M, the temperature difference decreasing between the two decades.

(c) At OWS K in the south, the difference increased by 0·2°C.

Kraus and Morrison (1966) compared the average daily variability of the air–sea temperature difference within calendar months with the variability due to differences between the same months in different years. Over monthly periods the fluctuations of the sea temperature were found to contribute very little to the variance of the air–sea temperature difference, and sea temperature variations became important only over longer periods.

Schell and Corkum (1976) have pointed out that, on account of differences between the thermal and dynamic characteristics of the ocean and atmosphere, sea temperature may continue to rise while air temperature is declining and may not 'catch up' with atmospheric temperatures for some time. This thermal lag during a period of climatic change does seem to have occurred in both the North Atlantic and the Mediterranean.

Air temperature over the oceans

A general parallelism exists between air and sea temperatures over periods longer than a few days. Constancy of air temperature is greater over the oceans than over the surrounding land masses, and over central parts of the North Atlantic Ocean monthly air temperature anomalies of more than 4°C are extremely rare. This is, of course, because cold air advected from polar regions, or in winter from the continents, becomes increasingly modified by its passage over the ocean.

There are latitudinal and seasonal changes in the average diurnal range of air temperature. The range is smallest in winter and greatest in summer; the observed increase of range with decreasing latitude reflects the influence of insolation. It is difficult, however, to study diurnal variations in high latitudes, since interdiurnal changes are not only small but are also largely masked by temperature changes associated with moving weather systems. The annual range of air temperature is maximum, 6°–10°C, between the parallels of 30°–40°N and declines to about 2°C near the equator. In mid-latitudes western parts of the oceans locally have ranges as large as 20°C. Prescott and Collins (1951) have shown that in eastern parts of the tropical Atlantic and Pacific Oceans air temperature maxima are reached more than 3 months after the date of peak solar radiation receipt.

The difference between two consecutive mean daily air temperatures is called the interdiurnal variability and Rosenthal (1960) shows that mean values over the North Atlantic Ocean range from 2·8°C in January at OWS D to only 0·4°C in July at OWS I and J. Obviously, advective changes are most frequent and intense in winter and air masses with source regions over the nearby cold continent arrive colder than the much-modified polar air.

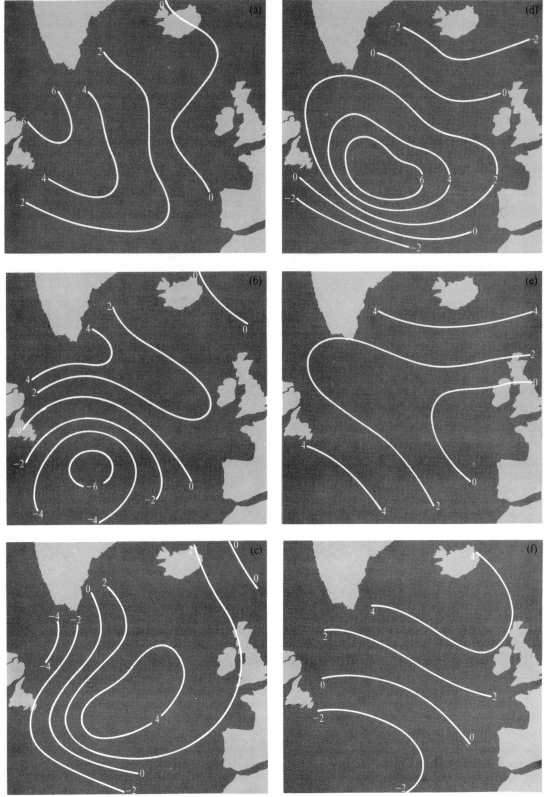

Fig. 5.6. First three eigenvectors of sea temperature anomaly fields in the North Atlantic in winter (a,b,c) and summer (d,e,f).

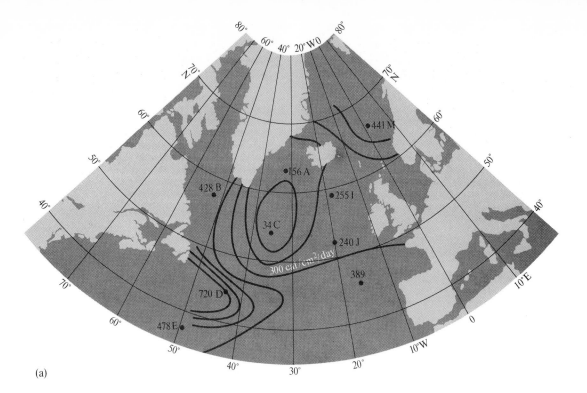

(a)

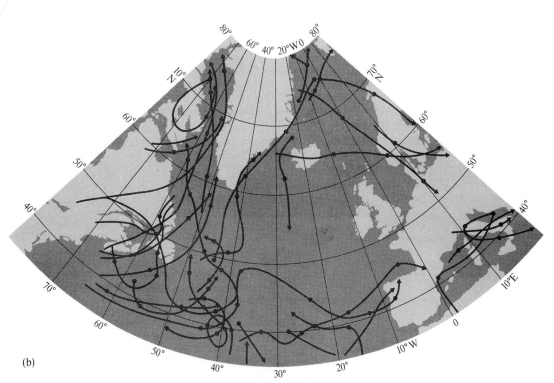

(b)

Fig. 5.7. (a) Turbulent fluxes over the North Atlantic Ocean February 1965. (b) Depression tracks over the North Atlantic Ocean February 1965.

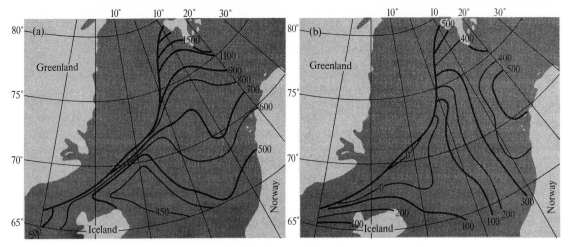

Fig. 5.8. Mean sensible heat flux over the Norwegian Sea in winter with (a) northerly airstreams, (b) southerly airstreams.

Similarly, in western parts of the Atlantic the North American land-mass is a more potent source of cold air at OWS D than polar regions. As windspeed increases the time taken by the air to reach southerly latitudes decreases, so that the time available for warming also decreases.

Synoptic heat budget studies

Changes of air temperature and SST occurring over the ocean on many different time-scales ensure that short-term fluctuations occur also in the individual terms of the energy budget, so that energy-transfer processes do not proceed continuously but in an interrupted, pulsating form. Global and hemispheric energy-budgets are considered in Chapter 4, but we need also to consider the magnitude of these short-term fluctuations. The work of Petterssen *et al.* (1962) and others has shown that the heat-exchange patterns are cellular and have a space-scale closely related to the major synoptic systems in the atmosphere. During months of anomalous atmospheric behaviour, anomalies will also occur in the pattern and intensity of the heat-budget terms. Vowinckel (1965) has noted that the overall expenditure and the relative importance of each term varied sharply in space from day-to-day during January 1963, and A. H. Perry (1968) has been able to demonstrate a relationship between the zones of maximum heat input to the atmosphere and the main depression tracks. In February 1965, for example, the input of energy seems to have been below normal over the entire North Atlantic. Even in south-western

parts (at OWS D and E) values were no more than average, whilst at OWS C there was the lowest air–sea energy exchange in any winter month between 1951 and 1966 (Fig. 5.7). Cyclonic activity was everywhere considerably below normal and depressions in the western Atlantic turned north up the Davis Strait or progressed eastward in low latitudes. There was also a corresponding tongue of relatively high air–sea energy input extending from the western Atlantic to Spain. It does seem possible to relate anomalous monthly circulation patterns to displacements in the distribution of the total turbulent heat flux over the North Atlantic Ocean. An investigation by White and Clarke (1975) of blocking ridge activity over the central North Pacific has shown that below-normal sensible-heat exchange was associated with blocking and that this anomalous heat flux arises in response to anomalous air-temperature fluctuations associated with the development of blocking ridges.

Heat-budget calculations carried out on the basis of well-defined circulation patterns have been attempted for the Norwegian Sea by Gagnon (1964). The sensible-heat flux was found to be the principal differentiator between types and in Fig. 5.8 the magnitude of the flux is shown for northerly and southerly airflow patterns. Near to the ice margin very high gains or losses of heat can ensue if a flow type is maintained over long periods, and this may significantly alter the ice limit and cover. Matheson (1967), working with data from the Gulf of St Lawrence, found that a severe winter will develop if the synoptic airflow in the autumn is such that it possesses the

117

capability of removing sufficient heat from the surface water layers to create an extensive ice-cover early in the season. The ultimate control is thus imposed upon the water body by the nature of the autumnal atmospheric circulation.

Lag-associations between North Atlantic sea temperatures and European atmospheric pressure

According to Pillsbury (1891), 'the moisture and varying temperature of the land depends upon the position of the currents of the ocean and it is thought that when we know the laws of the latter we will, with the aid of meteorology, be able to say to the farmer hundreds of miles distant from the sea, the winter will be cold and dry'. Later pioneering work was carried out by Helland-Hansen and Nansen (1920), at the Geophysical Institute in Bergen, but it is only in the last few years that the search for lag-associations between sea temperatures and subsequent atmospheric behaviour has been intensified. Ratcliffe and Murray (1970) have found that below-average sea surface temperatures over extensive areas to the south of Newfoundland are associated with blocked atmospheric patterns the following month over northern and western Europe, whereas when the ocean is warmer than usual in the same area a more progressive zonal type of circulation develops (see Fig. 5.9). In the latter case the development of Atlantic depressions is favoured, many of which then affect Europe. The explanation of such lag-associations undoubtedly lies in the complex anomalies in the heat budget that sea temperature anomalies induce. Extra water vapour transferred from the ocean to the atmosphere may be carried many hundreds of miles downwind in the upper flow before sensible heat is released by the condensation processes in the manner illustrated in Fig. 5.10. According to Lamb (1972), the thermal gradient across the Atlantic near $50°N$ is boosted when the sea near Newfoundland is warmer than average, but when the sea is colder than normal the gradient is strengthened near $30°N$ and cyclonic activity and the mainstream of the upper westerlies tend to be transferred to more southerly tracks, allowing large blocking anticyclones to be formed farther north over Britain and Scandinavia.

Longevity in the sea temperature anomaly pattern can lead to persistent or recurrent anomalous atmospheric circulation. Namias (1964) pointed out that a colder-than-normal ocean in the vicinity of OWS C during the period mid-1958 to early-1960 seems to have induced persistent blocking in the north-eastern Atlantic and negative precipitation anomalies over Scandinavia. During the summer of 1968 sea temperatures were as much as $4°C$ below normal near $44°N\ 50°W$ (Murray & Ratcliffe, 1969) and pressure during July–August was 8 mb above normal near the Shetland Islands. Concurrently, there was a pronounced anomalous east to north-east flow over much of Britain. Development of such cold water was apparently due to unusual cyclonic activity in the area at a critical time in the spring of 1968. This led to upwelling of cold water and a delay in the warming up of the ocean at a time when this heating is normally proceeding at its maximum rate. Four years later, in the summer of 1972, the melting of excessive amounts of pack ice in the western Atlantic led to cooling of the water and an enhanced north–south sea temperature gradient between this cold water and the warm North Atlantic current to the south (Perry, 1972). A strong EZ type of sea temperature pattern developed, and depressions originating over the Atlantic and intensifying along the zone of strong thermal gradient brought persistently dull and cool weather to Britain, causing June 1972 to be the coolest of the century at many places in the United Kingdom.

It has been observed by Oerlemans (1975) that a region with a positive or negative SST anomaly favours the occurrence of an upper ridge or trough, respectively, situated east of the anomaly. This may afford an explanation of the below-normal air temperatures and cyclonic types of weather that seem to occur frequently over the British Isles when sea temperatures are below normal in the eastern Atlantic (Perry, 1975). Two of the finest and driest summers in western Europe (1959 and 1975) have occurred when the entire western North Atlantic Ocean was colder than normal and the eastern half of the ocean warm. This tended to turn the zone of maximum thermal gradient from the normal W–E to a more SW–NE alignment, amplifying upper cold troughs over the western Atlantic and upper warm ridges over the eastern Atlantic and steering depressions away to the north of the British Isles.

Ratcliffe (1973) has suggested that lag-associations of use to the long-range weather forecaster may be found by studying sea temperature anomaly patterns in other specific areas. Table 5.3 suggests that a warm sea in the Biscay area in July favours warm weather in the following month over Britain.

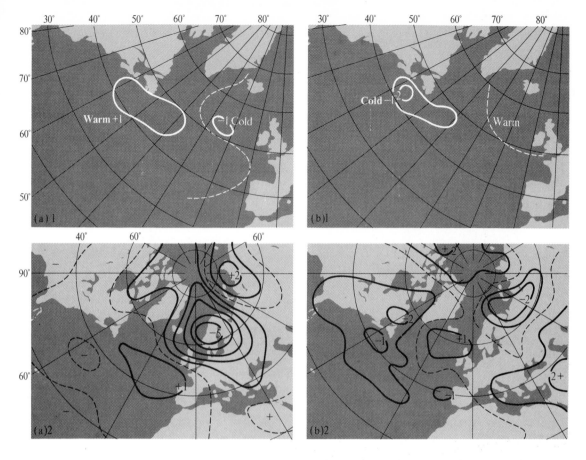

Fig. 5.9. (a) (1) Mean pattern of sea surface temperature anomaly (°C) in September in years with a positive anomoly off Newfoundland.
      (2) Mean sea-level pressure anomaly (mbs) in October for years corresponding to (1). Stippled areas represent values significant at 5 per cent.
  (b) (1) Mean pattern of sea surface temperature anomaly (°C) in December with a negative anomaly off Newfoundland.
      (2) Mean sea-level pressure anomaly (mbs) in January for years corresponding to (1). Stippled areas represent values significant at 5 per cent.

Table 5.3 Relationship between Biscay sea-surface temperature in July and British weather in August, after Ratcliffe, 1973.

Biscay sea temperature in July	Temperatures over England in August					England and Wales in August	
	Very cold	Cold	Average	Warm	Very warm	Dry	Wet
Warm (18 cases)	1	1	7	4	5		
Cold (18 cases)	8	3	2	2	3	7	4

119

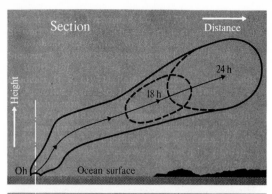

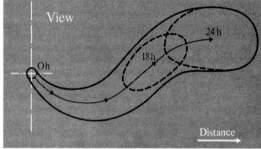

Fig. 5.10. Schematic presentation of the vertical and horizontal distribution of sensible and latent heat from a defined source at the sea surface.

Complex coupled air–sea systems in temperate latitudes

There is evidence that comparable mechanisms to the lag-associations described in the last section may be at work elsewhere. Namias (1969) has demonstrated a connection between the presence of anomalously warm water in part of the central Pacific and lower-than-usual pressure south of the Aleutian Islands in winter. These large-scale interactions between the atmosphere and the North Pacific may have been responsible for climatic fluctuations on time-scales of months and seasons and a space-scale larger than the ocean itself, perhaps hemispheric. For example, in the summer of 1962 warm water was generated near 40°N 170°W by persistent anticyclonic conditions. In the following autumn there was intense cyclogenesis in this area, as a consequence of the transmission of abnormal amounts of heat to the overlying atmosphere. Kraus (1972) has pointed out that it is difficult to understand the forced perpetuation of the initial warm sea-surface anomaly since the cyclonic activity might be expected to cause Ekman-layer divergence in the sea and therefore upwelling and cooling. Hence,

the stipulated feedback mechanisms which involve these processes are in his opinion necessarily highly speculative. The work of Namias indicates, however, that feedback conditions develop wherein the thermal state of the sea encourages the formation of atmospheric patterns which sustain its characteristic sea-surface temperature signature. The depressions stimulated to develop by the warm water lessen evaporative heat losses by dampening the lower air with rain and by reducing back-radiation through the formation of clouds.

The upper trough which became established over the central Pacific led to a response in the pattern of planetary waves downstream over North America and Europe, through the barotropic redistribution of vorticity. A ridge was favoured along the west coast of North America and a trough along the eastern seaboard of North America. These responses are due to *teleconnections* (see Bjerknes, 1969), through which the atmosphere transmits influences to distant areas. This pattern has recurred during other winters of the 1960s, as can be seen in Fig. 5.11, and Namias argues that a clue to winter circulations may often be obtained from the sea-surface temperature-pattern during autumn.

The spell of cold winters in Europe during the 1960s has been associated with the anomalous northerly and easterly wind components associated with the deep European trough, itself responsive to the upstream events over North America and the North Pacific. Developments in other areas often cooperated with and reinforced the instigated pattern, generating what Namias called 'positive feedback loops'. Farmer (1973) has confirmed that principal-component analysis of post-war sea-temperature data from the Pacific shows up the importance of anomalies in the central North Pacific and has been able to show that pressure is significantly higher over and to the north of Britain when Pacific sea temperatures are high. We should perhaps point out at this point, however, that Rowntree (1976a), using numerical techniques, has found that the exceptionally cold weather experienced in western Europe in the winter of 1962–3 was related to a cold SST anomaly in the tropical North *Atlantic* Ocean.

In a long series of papers Namias has given many examples of the effect of anomalous sea-temperature patterns, in both the Atlantic and the Pacific, on the overlying atmosphere, and a series of such relationships have now been established based on these case studies:

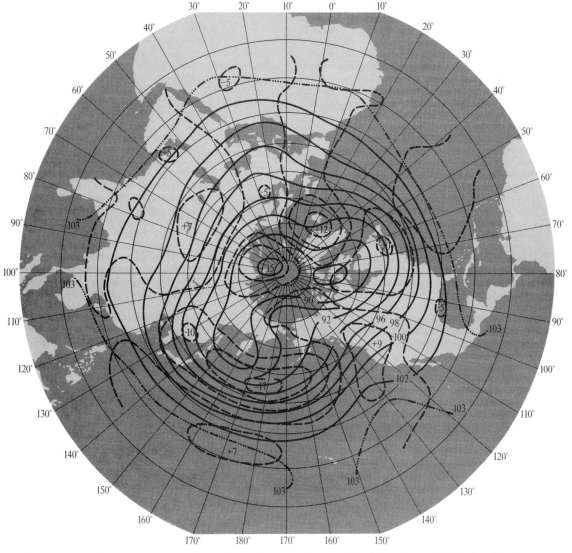

Fig. 5.11. 700 mb mean contours (solid lines labelled in tens of feet) and isopleths of anomalies (broken lines drawn for every 50 ft) for the seven winters 1961–7.

1. Warm water during summer off the east coast of North America can augment the supply of water vapour in southerly air-streams and increase rainfall, especially from tropical cyclonic storms. On the other hand, cold water along the Atlantic shelf region is associated with drought in north-eastern North America, as happened during the period 1962–6. Similar relationships have been found off southern Australia by Priestley (1964).

2. Cold pools of water generated during summer and early autumn in the Pacific have a tendency to be followed by anticyclogenesis over them, as in the autumns of 1961, 1966 and 1967. The water

temperatures often respond fairly rapidly to the abnormal anticyclogenesis, because more insolation and less upwelling occurs, and sea temperatures consequently rise. Thus, this autumn pattern is often self-defeating and usually short-lived.

3. Warm pools of water generated during summer and early autumn in the Pacific have a tendency to be followed by excess cyclonic activity over them. It must be stressed that in order for sea-temperature anomalies to affect the atmospheric circulation to any considerable degree the atmosphere must create the proper systems. Thus,

121

a warm pool of water in central latitudes of the eastern North Pacific during summer would not encourage cyclogenesis, because climatological constraints hardly ever place depressions over this area. Before recurrent cyclogenesis can occur the general circulation must be favourably disposed to steer embryonic depressions into areas where surface waters are abnormally warm.

4. Increased baroclinicity is generally established rapidly in the atmosphere over regions where strong oceanic temperature gradients exist, and this baroclinicity usually leads to increased cyclogenesis.

Among the most complex interactions that Namias (1974) has described is the coupled air–sea–continent system. The period from September 1972 to August 1973 demonstrated the remarkable continuity of a monthly-mean mid-tropospheric trough which moved from $155°E$ to $90°W$ between September and May. The development and motion of the trough over the ocean seems to be associated with and perhaps explained by sea-surface temperature contrasts supplying a baroclinic source of energy to the overlying atmosphere. Anomalously warm water was found ahead of the trough and cold water behind. Displacement computations of anomalous water-masses advected with normal surface currents indicated eastward motion of the cold and warm pools of about the same magnitude as the trough motion. After reaching the west coast of the United States in March 1973 the trough was supplied with a new source of baroclinic energy by the contrast between a cold polar Pacific air-mass, kept cold by persistent extensive snow-cover, and a warm tropical Gulf of Mexico air-mass, kept warm by seasonal heating over the southern plains of North America. Further motion of the trough eastwards was encouraged by an increase in wave-number and the generation of a new trough south of the Aleutian Islands. Regular migratory trough motions of this kind seem to occur fairly frequently, and Namias has discussed other examples, in particular those which occurred between summer 1957 and spring 1958 (Namias, 1959) and those which occurred in the autumn of 1965 (Namias, 1968).

Despite the prolific output of papers and reports from Namias, some basic questions remain unanswered, among them: how long a large atmospheric adjustment must last before the new flow pattern destroys or makes a substantial change in the SST pattern (see Namias, 1975); and

perhaps most important of all, why abrupt breaks in climatic regimes occur, such as occurred in the early 1970s, causing a quite different pattern of global climatic anomalies from those that characterized the 1960s.

Air–sea systems in tropical latitudes

Some of the most important work on very large-scale, almost global, air–sea interactions has been carried out by Jakob Bjerknes, who has found a link between equatorial sea-surface temperatures in the central and eastern Pacific Ocean, the extent and intensity of subtropical anticyclones and the extratropical Westerlies. He has used time-series of surface winds and sea-surface temperatures at Canton Island ($3°S\ 172°W$) to show that major temperature changes in this part of the Pacific are caused by varying strengths of the easterly winds which prevail there. When these are weak, as happened in late 1957 and early 1958 (Bjerknes, 1966), upwelling in the ocean is eliminated and sea temperatures rise above normal, while in the overlying atmosphere convection is stimulated and, as can be seen in Fig. 5.12, rainfall amounts increase.

The introduction of a large anomalous heat source also has the effect of intensifying the atmospheric circulation and in particular of allowing the Hadley circulation to transport northwards an excess of angular momentum, which in turn strengthens the zonal wind systems in the Northern Hemisphere. The expected increase in the westerly winds during the winter 1957–8 was observed in middle latitudes of the eastern Pacific, with a strong negative pressure-anomaly centred in the Gulf of Alaska. This, in turn, anchored the waves in the upper Westerlies in such a fashion that a positive pressure-anomaly became very persistent downwind (near south-west Greenland), causing a cold winter in northern Europe. Similar large-scale remote effects occurred in 1940–1 and 1952–3.

A version of the nine-level hemispheric model developed by the Geophysical Fluid Dynamics Laboratory of Princeton University, New Jersey (see Chapter 6), has been used by Rowntree (1972) to test Bjerknes' observations and hypotheses. The oceanic variations had important effects on the model atmosphere along the lines suggested; accordingly, it now seems that if ocean temperatures are monitored in the critical area useful atmospheric predications can be made.

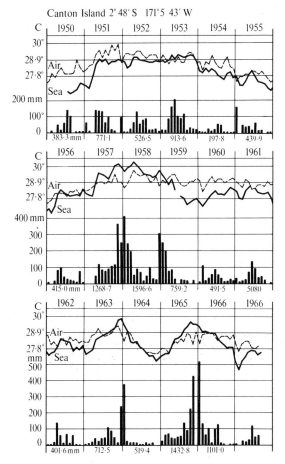

Canton Island 2° 48′ S 171°5 43′ W

Fig. 5.12. Time series 1950—67 of monthly sea and air temperature and monthly precipitation at Canton Island.

implications (see Caviedes, 1975). Large anomalies in the heat and water budgets of the ocean— atmosphere system cause the area-averaged precipitation amount (Quinn & Burt, 1972) to vary by a factor as large as 5 to 10. The immediate cause appears to be a general weakening of the south-east trade-wind circulation and an unusually southerly position of the Intertropical Convergence Zone.

Major El Niño events have occurred in 1891, 1925, 1941, 1957, 1965 and 1972. The extensive weather-satellite data that became available during the last occurrence has prompted several investigations of the phenomenon (see, for example: Wooster & Guillen, 1974; Ramage, 1975). Much remains unexplained, but Flohn (1973, 1975) has noted that the 1972 El Niño occurred simultaneously with drought in the Sahelian region of North Africa and deficiencies of monsoon rain in the Indian sub-continent. He has put forward a physical interpretation which is based on the concept of the wind-driven Ekman drift of the upper oceanic layer. Figure 5.13 suggests that in the Pacific Ocean three conditions are possible:

1. Equatorial upwelling of cold water caused by a symmetric wind-field, with two belts of convective activity, one on each side of the equator.

2. Equatorial upwelling and only one convergence zone, caused by an asymmetric wind-field.

3. Equatorial downwelling and a southward extension of the intertropical rainfall belt, causing high sea temperatures and strong convective rainfall (El Niño years).

It seems that transitions between the opposite conditions of 'cold and warm' equators occur rapidly and no intermediate stage can be maintained for more than a few months. Flohn (1973) has said 'the relatively short time of transition between two quasi-stationary opposing states seems to indicate a kind of vacillation or flip-flop mechanism'.

That the effects of El Niño are widespread is indicated by the study Namias (1973) made of hurricane Agnes, which passed across eastern North America in June 1972. This storm appears to have developed from meso-scale cloud systems which crossed the equator from their source-region, the abnormally warm Peruvian waters, and intensified when they encountered the favourable large-scale environment of the western Caribbean.

The far-reaching effects of SST anomalies in the eastern Pacific, which are being investigated by

Lamb (1972) has remarked that 'there seem to be a number of points at which small changes in the ocean or in the ocean—atmosphere relationship, could set up changes in the world climatic regime that would have a marked persistence or even a self-amplifying tendency'. Two such points may be mentioned. One is located at the branching of the Atlantic Equatorial current (at the 'nose' of Brazil) and the other in the waters off the Peruvian and Chilean coasts. In the latter area, off the west coast of South America, there is normally a widespread upwelling of cold water (see Chapter 3), but periodic interruptions of this upwelling take place. The name 'El Niño' is used to describe these occasions when large positive SST anomalies develop in the eastern South Pacific and heavy rainfall occurs as a consequence along the coasts of Peru and northern Chile. These El Niño events have important ecological and economic

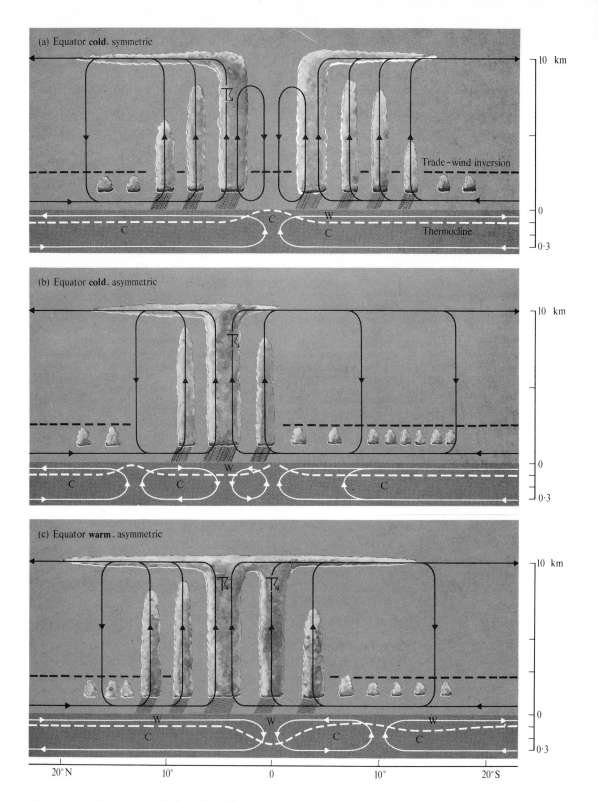

Fig. 5.13. Meridional atmospheric and oceanic circulation patterns and cloud distributions in the Eastern Pacific.

empirical studies and computer simulations, may become clearer as the Global Atmospheric Research programme progresses (see Chapter 6). As Flohn (1972) pointed out: 'the dynamic and energetic interaction between the ocean and the atmosphere forces oceanographers and meteorologists to talk and work together'.

Climatic change and the ocean—atmosphere system

The possible causes of climatic change can be divided into those which are external to the ocean—atmosphere—cryosphere[1] system, such as fluctuations in the solar emission or variations in Earth's orbital parameters, and those which are internal to the entire climatic system. Since climatic fluctuations extend over time scales from a few years to 10^8 years it is entirely possible that different causes may be at work. Lorenz (1968, 1970) is one of a growing number of scientists who have stressed that some scales of climatic change might be caused by natural fluctuations arising solely from the complex interactions that occur between land, oceans, atmosphere and polar ice. From what is contained in preceding chapters it is clear that the potential exists in these components to store or to release vast amounts of energy on time-scales ranging from days to centuries. Perhaps, as Namias (1972) has suggested, 'even the Ice Ages may be found to be grand manifestations of air—sea interactions'. If the climate of historical time has varied, in part, at least, as the result of quite spurious changes of the heat content of the oceans, there is certainly the possibility that the oceanic heat content may vary in a random fashion over long enough periods to play some part in the genesis of ice ages. There is also the possibility that as well as being a potential prime mover of fluctuations, air—sea interactions could be an important modifying influence on the character of fluctuations that are induced by other environmental stimuli (Mitchell, 1966).

It is conceivable that the atmospheric circulation can possess more than one state for any given set of boundary conditions, a condition which Lorenz (1968) has called 'almost intransitive'. There may not, therefore, be a unique set of atmospheric circulation statistics for a given set of boundary conditions, and as Schneider and Dickinson (1974) have said, 'we are faced with the possibility that for fixed external conditions widely different climatic states may be realized and it is possible

that almost-intransitivity could be interpreted by Man as a cause of observed climatic change'.

If we accept the postulate stemming from Faegri's (1950) ideas that the Pleistocene climatic variations were similar in kind to those of the historical period but simply persisted much longer, then it is essential to examine these more recent changes thoroughly. The oceans appear to have played a fundamental role in stabilizing the climate during the cold period of the Little Ice Age in the seventeenth and eighteenth centuries. Figure 5.14 shows the anomalies of sea temperature which prevailed between 1780 and 1820 (Lamb, 1972). It can be seen that positive anomalies prevailed south-west of Newfoundland, with negative anomalies in Icelandic and British waters. Depressions would find favourable conditions for formation and growth in the zone of strong thermal gradient south of the Newfoundland Banks and many of them would then move northwards into Baffin Bay and the Davis Strait. Certainly the average January pressure map for the period 1790–1829 sketched by Lamb and Johnson (1959) shows less cyclonic wind-stress south of Iceland compared with the present day (see Fig. 5.15). Physical reasoning has led Bjerknes (1965) to conclude that it is the northward-flowing geostrophic branches of the Gulf Stream that keep the Icelandic waters warm enough to be ice-free most of the year at the present time. All annual average pressure maps over the last century show cyclonic wind-stress in the north-east Atlantic and only for short periods have anticyclonic conditions south of Iceland temporarily opposed warm water advection in the North Atlantic Ocean. During the Little Ice Age the lack of geostrophic water transport would make the waters of the north-eastern Atlantic Ocean colder than at present, and such a negative SST anomaly, with its inherent lowering of the freezing-level in the precipitation-producing maritime airmasses, may have been a major factor in maintaining the Little Ice Age. Increased warm advection in north-western parts of the North Atlantic Ocean did not reach coastlines, but must have added to the moisture-content of south-easterly winds over Labrador and southern Greenland and led to enhanced snowfall and glacier-growth in these areas. Even if such reasoning is correct, the question remains as to how the Little Ice Age was initiated around AD 1600, as does the question of why the climatic anomalies did not grow by feedback into major ice-age proportions.

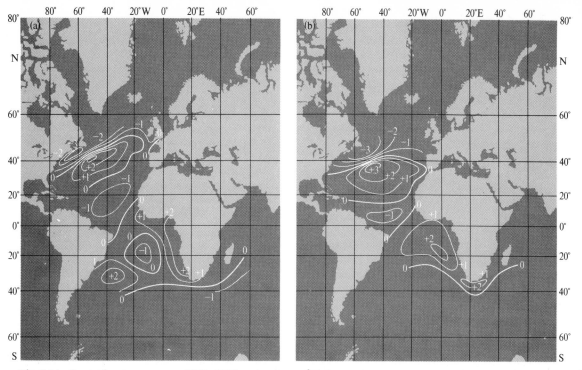

Fig. 5.14. Sea-surface temperatures 1780–1820 as departures (°C) from average 1887–1899 and 1921–38. (a) July. (b) January.

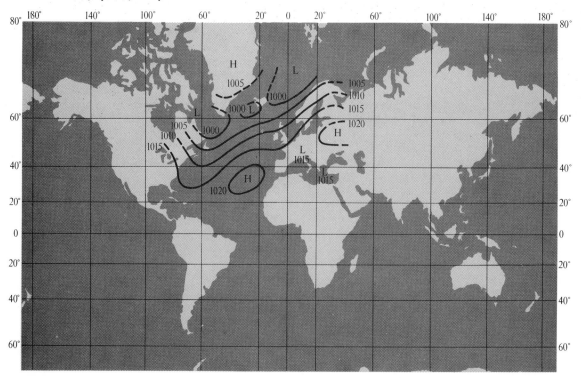

Fig. 5.15. Average January sea level pressure (mbs) for the period 1790–1829.

That a glacial climate may be initiated by reducing the salinity of the North Atlantic lies at the core of Weyl's (1968) theory of the rôle of the oceans in climatic change. At present, transport of water vapour across the isthmus of Central America contributes significantly towards maintaining a greater salinity in the North Atlantic than in the North Pacific. If this flux of water vapour were to be reduced and salinity in the North Atlantic thereby reduced, the main climatic consequence would be that the maximum ice limit in the Atlantic would be at a comparable latitude to that now found around the margins of the Pacific, with lower temperatures and extensive glaciation around the North Atlantic. It is interesting to note that the small salinity increase in the north-east Atlantic during the first few decades of this century did accompany a warming of the climate in the North Atlantic area and a decrease in sea-ice. Weyl points out that the salinity of the Atlantic controls the limit of growth of the Antarctic ice. This operates through the salinity and density characteristics of the Atlantic Deep Water circulating southward and coming up in latitudes south of the Antarctic Convergence. The crucial element in Weyl's theory, the atmospheric water–vapour flux across Central America, had to be estimated from the upper air soundings from 1956 to 1964 at just one station, Panama. During these years a reduction of the flux of about 30 per cent occurred, a trend which was in line with the significant weakening of the world's wind circulation over the same period. Thus, it does seem possible that changes in global wind patterns could affect the tropical easterlies across Panama. Once triggered into a glacial mode, the climate might return to inter-glacial conditions when stagnation of the bottom water has developed. Geothermal heating would then ultimately warm this water to the point where upward convection to the surface would start and salinity of the surface waters would increase again.

Large variations in amounts of ice and snow have been noted in both hemispheres during historical times. Sea-ice, particularly, represents a sensitive and vulnerable surface in a highly delicate state of balance. A change from ice to open water has a strong effect upon the heating of the overlying atmosphere, and, since thermal gradients are strong near the ice margin, any change in that margin will shift the gradients and associated depression tracks. At its greatest extent sea-ice in the Southern Hemisphere covers almost 6 per cent of the total oceanic area. Fluctuations in the

Antarctic sea-ice extent could perhaps affect global climate (Selby, 1973). Records indicate that in the past, and especially in the early nineteenth century, there have been many major ice-shelf surges (Radok et al., 1975) and even a relatively small surge can bring a series of interlinked consequences for the whole climate system.

With reference to the Arctic, Budyko (1966) has calculated that even a small increase (about 4°C) in summer temperature would cause the pack-ice to disappear in a few years, resulting in important changes in climatic conditions in the Northern Hemisphere. At present the pack-ice in summer severely inhibits storage of solar heat. Without pack-ice the most striking feature of the seasonal atmospheric cooling pattern would be the low rate of cooling in winter and the relatively high rate in summer. This, in turn, would strengthen meridional temperature gradients (and hence the atmospheric circulation) in summer and reduce them in winter (Fletcher, 1965).

Donn and Ewing (1968) have suggested that the Ice Age–Interglacial cycle can be explained by variations in the ice-cover of the Arctic Ocean. Their hypothesis uses as a starting point an ice-free Arctic Ocean, which would encourage evaporation. There would subsequently be condensation and precipitation; thus, ice sheets would be formed on the northern continents. Eventually, so much moisture would be locked up in the form of ice that sea-level would fall and the Arctic Ocean would no longer be supplied with Atlantic water over the Greenland–Scotland submarine ridge and the ocean would freeze over, denying further nourishment to the ice sheets, which would begin to melt and retreat. Schell (1971) has criticized the theory on the grounds that an ice-free Arctic would induce low pressure, with cold winds blowing polewards off the surrounding continents, so causing freezing of the ocean. Other palaeoclimatologists have pointed out that there is sedimentary evidence that the central Arctic has been ice-covered throughout the last 70,000 years.

The testing of fundamental hypotheses about climatic change is now becoming a reality with the advances in numerical modelling of the atmosphere that are taking place (see Barry, 1975). In the words of Schneider and Dickinson (1974): 'the models range from simple calculations of the effects of isolated processes on average conditions over the whole globe to space and time dependent computer simulations of a highly interactive system containing all known important processes'. Despite the progress of the numerical modellers, the empirical studies of environmental scientists will continue to be valuable, for, as Lorenz (1970) stated, 'It is inconceivable that in the near future we shall construct a model possessing every feature which could possibly be relevant. We therefore ought not to look upon a mathematical model as a means of by-passing the physical imagination needed to formulate hypothesis.' We consider ocean–atmosphere modelling in the next chapter; meanwhile, the words of Kraus (1972) offer a very apposite conclusion to this section:

one of the least well understood problems in environmental science, the explanation of climatic changes, offers a great challenge to students of air–sea interaction. The trouble is that we can conceive of such a variety of active feedback mechanisms — physical, chemical and biological — between the atmosphere and the oceans, and they all suffer from a deficiency of observational and theoretical constraints.

Note

1. 'Cryosphere' is the name used to describe regions of snow and ice.

Chapter 6

International projects and numerical models

As Chapman (1964) pointed out, the International Geophysical Year (IGY) of 1957–8 was by no means the first attack upon the great physical problems of the Earth. For centuries, indeed for at least two millenia, there have been endeavours to increase physical knowledge and understanding of our planet. Until the nineteenth century, however, geophysical research programmes tended to be uncoordinated, mainly individual, enterprises. The first large-scale project involving international scientific cooperation was the International Polar Year of 1882–3, when meteorological and other geophysical conditions in the Arctic were intensively observed over a period of 13 months. A second International Polar Year programme was completed in 1932–3, when, again, geophysical knowledge of the Arctic regions was much increased. On this occasion, however, there was also, simultaneously, a small observational programme in the Antarctic.

A proposal that a third International Polar Year project should take place in 1957–8 was adopted by the International Council of Scientific Unions (ICSU) in 1951, but the committee appointed to organize the venture decided to widen the scope of the project to embrace the whole Earth, and so the IGY was conceived[1]. In the event, nearly seventy nations, as well as various international scientific bodies, including the World Meteorological Organization (WMO), participated in the IGY enterprise. The cost of the exercise was shared by these nations and bodies. As we mention in Chapter 4, the collection of essays edited by Bates (1964) forms an excellent introduction to the IGY (and also to the International Quiet Sun Years (IQSY) of 1964–5, when there was a partial repetition of the IGY programme).

Meanwhile, geophysicists were beginning to succeed in their attempts to simulate numerically the behaviour of the ocean–atmosphere system, and, in 1956, N. A. Phillips published an account of what he called 'a numerical experiment' on a model two-layer dry atmosphere extending from polar to near-equatorial latitudes. In the words of Sheppard (1968):

He made it start from rest at a uniform temperature but gave it more or less realistic sources and sinks of heat and computed its evolution. After 'jolting it', he found it developed disturbances, with structures much like our familiar cyclones and anticyclones, which travelled in a belt of westerlies with surface easterlies on their warm equatorward and cold poleward sides. **It was the first demonstration of the necessity for something very like the observed general circulation of the atmosphere.**

The success of Phillips' work stimulated J. Smagorinsky and Y. Mintz, among others, to further investigate the possibilities of modelling mathematically atmospheric and oceanic circulations (see Lorenz, 1967; Smagorinsky, 1970). We discuss theoretical models later in this chapter, but we must record at this juncture that developments in geophysical modelling, together with developments in satellite technology and computer capabilities have precipitated a further great international geophysical enterprise, the Global Atmospheric Research Programme (GARP), which we discuss at length in this chapter.

Before we proceed to discuss GARP, however, we recall that reference is made in Chapter 2 to another major collaborative venture, the International Indian Ocean Expedition (IIOE) of 1959–65 (see also Ramage, 1965). We also mention that several international organizations besides the WMO have marine research within their terms of reference. The IIOE, for example, was sponsored by the Scientific Committee on Oceanic Research (SCOR) and the United Nations Educational, Scientific and Cultural Organization (UNESCO) and was coordinated by the Intergovernmental Oceanographic Commission (IOC), a body established in 1960. To quote Groen (1967), the IOC was 'the first world-wide organization covering the whole domain of oceanology [2] . Among its accomplishments have been the inauguration of the Long Term and Expanded Programme of Oceanic Exploration and Research (LEPOR) and the International Decade of Ocean Exploration (IDOE). For details of LEPOR, IDOE and most marine science activities involving international cooperation reference should be made to the relevant WMO publications, in particular, the quarterly *WMO Bulletin*. We now turn to GARP, an enterprise which has lately excited great fervour among students of the ocean—atmosphere system and which is likely to prove a landmark in the history of endeavours to understand this system.

The Global Atmospheric Research Programme

Whereas the chief objective of the IGY (and also of the IQSY) was, to quote Bolin (1970), 'to obtain an enriched worldwide complex set of observational data', GARP, quoting Bolin again, 'is a global *research* programme, requiring a coordinated effort in which quite a number of researchers, research institutions and national organizations will contribute in their own way to solving the scientific, technological, operational and logistic problems involved in various phases of the programme'.

At its first meeting, in April 1968, the Joint Organizing Committee of GARP (JOC) defined precisely the objectives of the project in these words (see Garcia, 1968):

GARP is a programme for studying those physical processes in the troposphere and stratosphere that are essential for an understanding of:

(a) the transient behaviour of the atmosphere as manifested in the large-scale fluctuations which control changes of the weather; this would lead to increasing the accuracy of forecasting over periods from one day to several weeks;

(b) the factors that determine the statistical properties of the general circulation of the atmosphere which would lead to better understanding of the physical basis of climate.

This programme consists of two distinct parts, which are, however, closely interrelated:

(i) the design and testing by computational methods of a series of theoretical models of relevant aspects of the atmosphere's behaviour to permit an increasingly precise description of the significant physical processes and their interactions;

(ii) observational and experimental studies of the atmosphere to provide the data required for the design of such theoretical models and the testing of their validity.

It was decided further at this meeting that GARP would be composed of a series of *Subprogrammes,* comprising such projects of both a theoretical and an experimental character as would be consistent with GARP objectives, and *Experiments,* comprising large observational programmes for ascertaining the behaviour of the whole atmosphere or some part of it, depending upon the requirements of a particular subprogramme.

The first major observational experiment in GARP was the Atlantic Tropical Experiment (GATE), which (as we mention in Chapter 4) took place during the summer of 1974. The objectives of this experiment were 'to extend our knowledge of those aspects of the meteorology of the equatorial belt that are essential for a proper understanding of the circulation of the atmosphere as a whole and, at the same time, to improve the understanding and prediction of weather in the tropics' (Mason, 1973). The vast tropical and equatorial oceans constitute the source regions for a large proportion of the energy required to drive atmospheric and oceanic motions. Thus, as Garstang *et al.* (1970) have noted, 'investigations into planetary energetics must logically place considerable emphasis on these tropical ocean areas, from which the absorbed solar energy is returned to the atmosphere across the air—sea boundary surface'.

Various elements of the ocean—atmosphere system were studied during GATE:

1. The characteristics of atmospheric wave disturbances, phenomena typically a few thousands of kilometres in length (see, for example, Kuettner, 1974). These were investigated with the aid of enhanced synoptic upper-air networks over Africa and South America, aerological soundings from dedicated ships on the Atlantic Ocean and observations from a geostationary satellite, the object being that the origin, structure, development, propagation and energy sources of the waves, as well as momentum and heat transfers in them, might become better understood. The relationships between easterly waves and cloud clusters embedded within them were of particular interest.

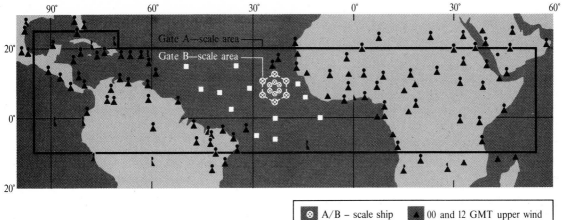

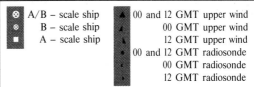

Fig. 6.1. The distribution of observing ships and land-based upper-air stations in the GATE A- and B-scale areas.

2. The internal structure of cloud clusters (which are typically 100 to 1,000 km in diameter) and the interactions of these clusters with larger-scale atmospheric systems were investigated, with the aid of a special mesoscale network of observing ships and instrumented aircraft in a 1,000 × 1,000 km polygon located about 1,000 km off the west coast of Africa (the B-scale area in Fig. 6.1). Upper-air soundings of temperature and wind were made from many of the ships, and radar was used for ascertaining precipitation patterns.

3. Fluxes of sensible heat, water vapour and momentum associated with mesoscale convective systems, individual cumulus cells and atmospheric phenomena of even smaller scale, were measured by means of a 'Boundary-layer Instrument System' which involved the use of tethered balloons, special sondes, buoys and gust-probe aircraft.

4. Condensation and freezing nuclei were sampled and their concentrations measured at various levels in the atmosphere.

5. Sea-surface temperatures and the dimensions of wind waves and swell were measured from ships and aircraft. Additionally, an international group of oceanographers concentrated upon the upper ocean, particularly upon its response to impressed atmospheric disturbances of various scales. A special study was made of the Equatorial Undercurrent (the Lomonosov Current).

Although some observational difficulties were encountered — for example: a few stations did not become operational and some equipment failed to function satisfactorily — the field phase of GATE was undoubtedly a success (see Kuettner & Parker, 1976), and MWS (1975) feels 'there is good reason to believe that the data collected during GATE will prove adequate for achieving the basic objective of the experiment'. Not until the data have been processed, however, will it be possible to judge whether or not the scientific objectives of the project have been met. Certainly, though, valuable new knowledge was obtained, perhaps the most startling discovery being that cloud clusters of very large size can develop extremely rapidly ('in a matter of less than six hours', according to MWS).

Plans are now well advanced for the First GARP Global Experiment (FGGE), which is scheduled for 1977–9. It is intended (see WMO, 1976) that this experiment should consist of a preparatory phase, from September 1977 to August 1978, during which there will be an intense observational programme (using the synoptic and upper-air network established for the World Weather Watch[3] supplemented by specially-equipped ships and aircraft, satellites, buoys and constant-level balloons) and an operational phase, from September 1978 to August 1979. Nitta (1975) has written: 'One of the major objectives of the FGGE is to develop, test and employ more powerful methods for assimilating meteorological observations, into global numerical models.' He pointed out that definitive global data sets need to be constructed if this aim is to be achieved, and he gave the following essential functions of the data sets:

1. to provide initial and verifying conditions for experiments designed to extend predictability in the medium range (from a few days to a few weeks) and in the extended or long range (from a few weeks to several months);

2. to define the necessary elements of an operational global observing system with a reliable assessment of the consequences for predictability of various compromises;

3. to test systematically understanding of atmospheric dynamics through parameterization hypothesis and as a result of computational resolution;

4. to define diagnostically the structure, variability, energetics and transport mechanisms of the general circulation for the particular year of the FGGE, thereby providing the basis for interactive studies between, for example, the stratosphere and troposphere, the atmosphere and the ocean, the tropics and the extra-tropical regions, and the hemispheric circulations; and

5. to provide information essential to progress towards meeting the second GARP objective of understanding the physical basis of climate.

An ambitious project indeed!

During the formative stages of GARP various influential committees stressed the importance of the oceans in the genesis and maintenance of atmospheric circulations. For example, a WMO Working Party noted at a meeting in 1969 (see WMO Secretariat, 1970): 'The air and the sea together form a vast heat engine on which our weather and climate largely depend, and a major aim is the development of an ocean–atmosphere prediction model of global application.' Nevertheless, there was for several years surprisingly little emphasis in GARP upon the concept of an ocean–atmosphere *system*. However, it is evident from the following passage (WMO, 1976) that attempts are now being made to integrate atmospheric and oceanic studies:

Because of the growing awareness of the nature of the coupling between the ocean and the atmosphere, oceanographers and meteorologists have begun to consider how to plan joint observational and research programmes in areas of common interest.

FGGE offers not only the possibility of a comprehensive data set for atmospheric scientists, but it also provides an unparalleled opportunity for oceanographers to study this coupling in detail.

The recent institution of a GARP Climate Dynamics Sub-programme is further evidence that the necessity to consider atmospheric and oceanic behaviour in terms of a fluid system is now well appreciated.

At its meeting in the summer of 1974 the JOC recommended that this sub-programme should concentrate upon processes associated with variations of the ocean–atmosphere system over periods of time ranging from about a month to the order of a century (MJR, 1975). In October 1975 the JOC defined the purpose of the project in much greater detail (see WMO, 1976). It was decided that a distinction should be made between the *internal climatic system* (comprising the atmosphere and the oceans and such features of the land surface as are required to describe the hydrological cycle) and the *external system* (consisting of various forcing factors). It was recognized, quoting WMO (1976), that, 'ultimately the dynamics of the internal system could only be well understood on the basis of *comprehensive coupled ocean–atmosphere models* but it was realized that dynamical modelling of atmospheric motions and oceanic circulations could not proceed at the same pace.'

The JOC considered three approaches to the modelling problem appropriate:

1. atmospheric dynamics and thermodynamics should be studied separately, taking the sea-surface temperature as given or as observed;

2. oceanic dynamics should be studied separately, taking meteorological parameters such as wind stress and cloudiness as given or as observed; and

3. the problems of coupling atmospheric and oceanic models, arising in particular from the very different characteristic time and space scales of the two media, should be studied without waiting for the difficulties in modelling, either of the atmosphere or of the ocean separately, to be fully resolved.

Quoting further from WMO (1976):

Whilst the modelling of the atmosphere using comprehensive three-dimensional general circulation models is feasible, it is not so for the ocean, the dynamics of which are much less known. It was therefore decided to institute a global ocean modelling study which would require a broad programme of numerical experimentation directed primarily at the clarification of the various interactions which must ultimately be parameterized in a global oceanic model. In particular, the numerical studies should be carried out for three reasons: first, to investigate the dynamics of quasigeostrophic eddies and the coupling between ocean circulation regimes of different depths and response times; second, to study the response of the ocean to variable atmospheric forcing in the Equatorial and upwelling regions; and third, the response of the atmosphere to sea-surface temperature anomalies. The sensitivity of the models to the parameterization of sub-eddy scale processes, in particular internal gravity waves and seasonal sea-ice predictions based on various parameterizations of sea-ice formation, should also be tested.

It is hoped that parameterization schemes based upon data collected during the Air-Mass Transformation Experiment (AMTEX) will prove successful for representing adequately air-mass modification in global atmospheric models. The field phases of this experiment, which took place in February 1974 and February 1975 near the south-western islands of Japan, concentrated upon transfer processes in cold airstreams flowing across the warm Kuroshio Current (see Lenschow, 1972, 1974). The large input of sensible and latent heat to the atmosphere over this current in winter months (see Chapter 4) results in the development of atmospheric phenomena ranging in scale from individual cumulus clouds to planetary waves more than 3,000 km in length. In the scientific programme of AMTEX there was, accordingly, some emphasis upon:

(a) fluxes of sensible heat, water vapour and momentum in the atmospheric boundary layer;

(b) vertical transfers of heat and momentum from boundary layer to the overlying free atmosphere by means of cumulus convection; and

(c) relationships between meso-scale and synoptic-scale disturbances, with particular reference to the amount of energy supplied from the ocean surface during air-mass modification.

In the Royal Society Joint Air–Sea Interaction Project (JASIN), a contribution by the United Kingdom to GARP, there is to be intensive study of the atmospheric boundary layer and the upper ocean in an area of the North Atlantic Ocean about

450 km west of the Outer Hebrides. Buoys, ships, balloons, aircraft and satellites are to be used in the observational programme, which will concentrate upon the structures of these turbulent layers and upon interactions of the layers with larger-scale features of the ocean–atmosphere system. The primary aims of JASIN are (Royal Society, 1975):

1. to observe and distinguish between the physical processes causing mixing in the atmospheric and oceanic boundary layers and relate them to the mean properties of the layers; and

2. to examine and quantify aspects of the momentum and heat budgets in the atmospheric and oceanic boundary layers and the fluxes across and between them.

Instrument trials for the JASIN project were carried out near Ocean Weather Station 'Juliet' (about $52°N$ $20°W$) in June 1970 and September 1972. The principal field phases of the project are scheduled to take place near Rockall, in an equilaterally triangular area of side 200 km, between mid-August and mid-September 1977 and between mid-July and mid-September 1978.

Since the inception of GARP there have been other major boundary-layer investigations, in particular:

(a) the Atlantic Trade-wind Experiment (ATEX), which took place in the period 6 to 21 February 1969 and concentrated upon an area of the mid-Atlantic near $10°N$ (see, for example, Augstein *et al.*, 1974); and

(b) the Barbados Oceanographic and Meteorological Experiment (BOMEX), which took place in the summer of 1969 in a 500 × 500 km square near Barbados (see, for example: Davidson, 1968; Garstang *et al.*, 1970; Holland, 1972).

The Complex Atmospheric Energetics Experiment (CAENEX), which took place in the Soviet Union between October 1970 and March 1972, was mainly devoted to a study of atmospheric processes occurring over land surfaces (see Kondratyev *et al.*, 1970, 1973).

We may also mention at this point a major air–sea interaction project which is independent of GARP, namely the North Pacific Experiment (NORPAX). This long-term experiment (1972–82) is part of the IDOE programme and is supported by the US Office of Naval Research (ONR) and the National Science Foundation (NSF). Various American research groups are participating in the enterprise, principal among them being a group from the Scripps Institution of Oceanography, San Diego, California. The organizers of NORPAX hope that the project will eventually attract scientists of all nations.

NORPAX grew out of an attempt by scientists of the Scripps Institution to test the postulate that relationships exist between sea-surface temperature (SST) anomalies in the North Pacific Ocean and climatic events over eastern parts of this ocean and the continent of North America. When the inadequacy of the initial research effort to solve a problem of such complexity became obvious, support for a larger, more comprehensive, research programme was sought. Resources made available by NSF were combined with those of ONR, who supported the original study, and provision for NORPAX was thus assured.

The major objective of NORPAX, in the words of Barnett (1973) and Collins *et al.* (1973), is: 'to describe and develop a basis for explaining the mechanisms responsible for the large-scale oceanic and atmospheric fluctuations that occur in the mid-latitudes of the North Pacific Ocean'. It is believed that the fluctuations might be explained in terms of interactions between and within major atmospheric and oceanic phenomena, since all contiguous elements of the ocean—atmosphere system interact with each other to some degree. Accordingly, research endeavours are aimed at elucidating these interactions and so providing a sound physical basis for long-range weather prediction.

The first field experiment of NORPAX, a project called POLE, took place during January and February 1974 about 1,440 km north of Hawaii. According to an Annual Report of the Scripps Institution (Scripps, 1975):

This experiment was the first in a series that will eventually culminate in a heat-budget study aimed at determining the mechanics of SST anomaly generation . . . [and it] . . . had the modest goals of determining the principal space/time scales of atmospheric flow-patterns and sea-surface temperature-patterns.

The experiment was named POLE to indicate that studies were confined to a small horizontal area but extended some way vertically through the upper ocean and overlying atmosphere; that is, columns of air and water of relatively small cross-sectional area were examined.

Returning to the subject of GARP, we now mention another of its sub-programmes, The

Monsoon Experiment (MONEX), which is scheduled to begin in 1977. During this experiment there is to be intensive study of (a) wind and temperature profiles in the lower and middle troposphere over the Arabian Sea and (b) circulations in monsoon depressions and mid-tropospheric disturbances over the Bay of Bengal and meso-scale convective systems over the China Sea. Particular attention will be paid to the factors thought to determine the onset of the south-west monsoon and to the atmospheric conditions associated with active periods and breaks in the monsoon. There will also be study of winter circulations over southern parts of Asia and over adjacent oceans.

Although it was originally intended that MONEX should concentrate exclusively upon the monsoon of southern Asia, it now seems likely that its scope will be widened somewhat by the inclusion of a subsidiary experiment, for it was reported in WMO (1976) that the JOC felt at its meeting in October 1975 that 'a study of the West African monsoon could be very profitably mounted by West African countries during FGGE so that advantage could be taken of the enhanced observational coverage at that time'.

Finally in this catalogue of GARP experiments and sub-programmes concerned directly with interplay between the oceans and the atmosphere, we draw attention to The Polar Experiment (POLEX), the objectives of which are as follows (Kellogg, 1974):

1. the achievement of a better understanding of energy transfer processes and heat budgets in the Arctic and Antarctic, so that these regions might be properly represented in numerical models of the atmospheric general circulation; and

2. the provision of adequate data from polar regions during the FGGE.

More specifically, it is envisaged that POLEX, a long-term project extending over the period 1973—83, will embrace the following components:

POLEX (NORTH)
Meteorological observations required for weather forecasting;
Physical processes at and above the ice;
Heat balance of the Arctic Ocean and adjacent seas;
Numerical modelling experiments;
Establishment of a Polar Climate Data Centre.

POLEX (SOUTH)
Meteorological observations for FGGE;

Numerical modelling of a coupled ice–ocean–
    atmosphere system;
Parameterization of the polar planetary boundary
    layer;
Climatic record for the Southern Hemisphere;
Dynamics and thermodynamics of the Southern
    Ocean (all time-scales);
Dynamics and thermodynamics of Antarctic sea-ice.

The Arctic Ice Dynamics Joint Experiment
(AIDJEX) has already proved to be a valuable
contribution to POLEX, for its field phase, which
took place on the Beaufort Sea in 1975 and 1976,
has yielded much new information about sea-ice,
particularly about its morphology. The objective
of AIDJEX was (National Academy of Sciences,
1974): 'to reach, through field experiments and
theoretical analyses, a fundamental understanding
of the dynamic and thermodynamic interactions
between Arctic sea-ice and its local environment'.
Those wishing to learn more of this experiment
may turn profitably to the series of AIDJEX
Bulletins, which are published by the Division of
Marine Resources of the University of Washington,
Seattle. The history of POLEX and a summary of
the aspirations of the Soviet scientists who
proposed the project have been given by Weller
and Bierly (1973).

We conclude our discussion of GARP and
related projects by noting that international
research enterprises are extremely expensive, and
those who contribute to them financially expect
some return for their money. It should always be
borne in mind, especially by persons directly
involved in the enterprise, that the money devoted
to increasing knowledge and understanding of the
ocean–atmosphere system is *invested* for the
eventual benefits of mankind. The enterprises are
more than intellectual exercises. Professor
B. R. Bolin (1970), the first chairman of the JOC,
put it thus:

*The atmosphere is an important part of the human
environment. We cannot deal with the global
atmospheric pollution problems unless we know
the behaviour of the atmosphere well enough and
we cannot say anything about their influence on
weather and climate, unless a programme of the
kind outlined in GARP is realized. GARP should
be placed in this very much wider context. At the
same time the objectives and the means of realizing
them must be safeguarded. The integrity of the
programme must be maintained, not the least in
order to keep costs under control. The experiments
proposed must be well-defined and not imply open-
ended commitments by Governments willing to
contribute to their implementation.*

Some advanced numerical models of climate

It is clear from the foregoing that there is
nowadays a very considerable emphasis upon
numerical modelling of atmospheric and oceanic
phenomena and interactions between them. As
Spar and Atlas (1975) have noted: 'the complete
physical consequences of interactions between the
sea and the air can ultimately be calculated only
through the use of coupled air–sea models in
which each fluid is free to respond to the influence
of the other'.

The most advanced interactive model of the
ocean–atmosphere system is that developed at the
Geophysical Fluid Dynamics Laboratory (GFDL)
of Princeton University, New Jersey, where for
more than two decades dynamicists have
concerned themselves with the problems of
modelling fluid systems mathematically. Until
about 10 years ago research groups associated with
GFDL (formerly known as the Institute for
Advanced Study) developed separately numerical
models of atmospheric and oceanic circulations. A
group led by Joseph Smagorinsky extended the
work of Norman Phillips (1956) on models of the
atmospheric general circulation (see Smagorinsky
et al., 1965; Manabe *et al.*, 1965), and a group led
by Kirk Bryan modelled the oceanic general
circulation (see Bryan & Cox, 1968; Bryan, 1969a).
In due course the groups combined their expertise
and addressed themselves to the task of
constructing a realistic model of the ocean–
atmosphere system.

The first simulation of climate with a combined
ocean–atmosphere model was that discussed by
Manabe (1969a, b) and Bryan (1969b). A nine-level
atmospheric model, similar to that described by
Manabe *et al.* (1965), was used to calculate values
of atmospheric variables at each point of a
horizontal grid on which points were spaced
approximately 500 km apart, and a five-level
oceanic model, similar to that described by Bryan
and Cox (1968), was used to calculate values of
temperature, salinity and current-velocity at each
point of a horizontal grid on which points were
spaced approximately six degrees of latitude and
three degrees of longitude apart (except near the
western boundary of the model ocean, where a
spacing of less than one degree of longitude was

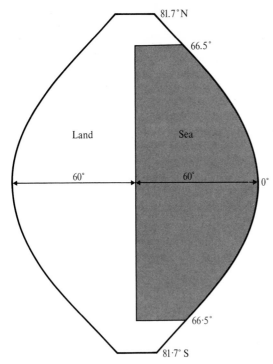

Fig. 6.2. The ocean—continent distribution chosen for the pioneering ocean—atmosphere numerical model developed at the Princeton Geophysical Fluid Dynamics Laboratory.

used). To facilitate interpretation of results the simulation work was carried out in three stages:

1. The atmospheric model was studied with the ocean acting as an infinite reservoir of moisture for the atmosphere and heat transfer by ocean currents neglected.

2. The oceanic model was studied on the assumptions that there is no feedback from the atmosphere, and that the distributions of surface temperature, wind-stress and precipitation are constant with respect to time; and

3. The atmospheric and oceanic models were allowed to interact fully with each other.

Manabe and Bryan (1969) considered the most interesting result of their work to be 'the quantitative demonstration of the effect of ocean currents on the distribution of temperature, relative humidity, and precipitation patterns'.

In this pioneering model of the ocean—atmosphere system a very simple distribution of land and sea was assumed and, to save computation time, calculations were carried out for a region spanning only 120 degrees of longitude (see Fig. 6.2). Later versions of the model were more

realistic (see, for example: Manabe *et al.*, 1975; Manabe & Holloway, 1975; Bryan *et al.*, 1975).

Using the GFDL model of the atmospheric general circulation it is now possible to investigate in some detail specific problems, including, of course, problems concerning interaction between atmospheric and oceanic phenomena. The study made by Manabe *et al.* (1974) of 'the seasonal variation of the tropical circulation as simulated by a global model of the atmosphere' is an example of such an investigation.

Manabe *et al.* (1974) found that the model accurately simulates seasonal patterns of rainfall in the tropics and associated wind-fields. It also produces precipitation systems which average about 1,000 km in diameter and resemble cloud clusters (Fig. 6.3). Moreover, the model indicates that, over the sea, the locations of rainbelts and accompanying tropical disturbances are determined primarily by distributions of SST. This being so, Manabe *et al.* felt that relationships between SST anomalies and tropical precipitation patterns should be investigated numerically.

Shukla (1975) has responded to this suggestion and demonstrated with the GFDL model that the effect of introducing a cold SST anomaly into the Arabian Sea near the Somali Coast is to drastically reduce rainfall over India (see Fig. 6.4).

Numerical models of the atmosphere other than the GFDL model have also been used in studies of SST anomalies. For example, the six-layer model developed at the US National Center for Atmospheric Research (NCAR) has been used by Schneider and Washington (1973) to investigate possible relationships between global SST anomalies and world cloudiness. This study indicated that a positive anomaly of $2°C$ would decrease amounts of low cloud by $2·74$ per cent and a negative anomaly of the same magnitude would increase amounts by $1·53$ per cent. Both amounts were found to be statistically significant.

Houghton *et al.* (1974), in contrast, used the NCAR model to investigate the effects of SST anomalies off the coast of Newfoundland on the frequencies, tracks and intensities of depressions over the North Atlantic Ocean. Numerically-predicted responses of the atmosphere to SST anomalies compared well with those observed, particularly on occasions of warm anomalies. Huang (1975) has also used the NCAR model to study mid-latitude atmospheric responses to oceanic forcing (by the North Pacific Ocean) and has shown that warm SST anomalies encourage cyclogenesis. Forcing of atmospheric motions by

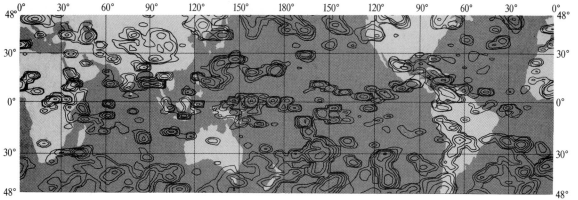

Fig. 6.3. The mean rate of precipitation over a period of six hours on 14 July, according to the GFDL numerical model. Contours are shown for: $0 \cdot 1$, $0 \cdot 2$, $0 \cdot 5$, $1 \cdot 0$, $2 \cdot 0$, $5 \cdot 0$, $10 \cdot 0$, and $20 \cdot 0$ cm day^{-1}.

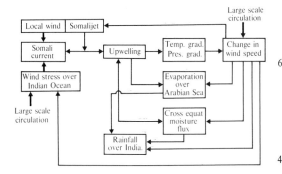

Fig. 6.4. Schematic representation of ocean–atmosphere interactions over the Arabian Sea, according to Shukla (1975).

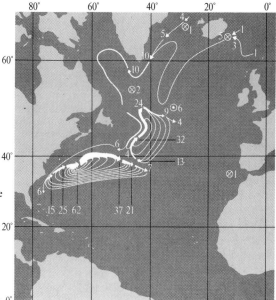

Fig. 6.5. Circulation of deep water ($T < 4°C$) in the North Atlantic Ocean. The numbers indicate water transports in units of 10^6 m sec^{-1}.

tropical SST anomalies has also been studied by Rowntree (1976b), using the five-level model of the United Kingdom Meteorological Office.

A nine-level model of the general circulation of the atmosphere developed at the Goddard Institute for Space Studies (GISS) has been used by Spar and Atlas (1975) in a weather-prediction experiment involving comparison of observed and predicted atmospheric responses to SST anomalies prevailing over the Northern Hemisphere on 20 December 1972. These initial conditions included widespread positive anomalies in the North Pacific Ocean, in some places as much as $4°C$, and negative anomalies of $3°C$ near Newfoundland. The GISS model recognizes that convective precipitation is enhanced over moderately large positive anomalies and suppressed over negative anomalies. Nevertheless, it was found that the inclusion of observed SST anomalies in the model did not significantly improve forecasting skill.

The global response of the atmosphere to a persistent warm anomaly in the surface waters of the extra-tropical North Pacific Ocean has been investigated numerically by Spar (1973). He found that the meteorological effects of the anomaly spread polewards, with a stronger response in winter than in summer, and, surprisingly, that after a simulation-month of thermal forcing the atmospheric response to an SST anomaly is at least as large in the opposite hemisphere as in the hemisphere of the anomaly. He concluded: 'clearly the atmospheric response as represented by the model computations is a complex function of the

137

initial state of the atmosphere and the hemispheric topography on which the anomaly is superimposed and could not have been anticipated from simplistic qualitative reasoning'.

In a further experiment Spar examined a transient anomaly which was allowed to persist for only one simulation-month while the model was allowed to 'run' for three simulation-months. The effects on computed monthly-mean sea-level pressure fields over a season were found to be as large in absolute magnitude as those generated in the model by a persistent SST anomaly. When interpreting these results, however, Spar sounded a note of caution, pointing out that the effects of random errors in the model may completely overwhelm any systematic effects due to anomalies. Nevertheless, he considered 'there is no reason to doubt that the behaviour of the model may indeed reflect a similar sensitivity of the real atmosphere to the temperature of the sea-surface'.

The *mechanics* of ocean–atmosphere coupling has been investigated by Pedlosky (1975), using a simplified but non-linear model which involves interaction between a finite-amplitude cyclone wave, the large-scale atmospheric temperature field and sea-surface temperatures. In particular, Pedlosky sought to ascertain 'the process which maintains and even intensifies an oceanic sea-surface temperature anomaly'; a matter he considered 'a key conceptual question at the heart of the air–sea coupling problem'. He made good progress towards this objective, showing that, 'when there is a large heat release to the atmosphere and long-term storage of heat in the mixed layer', embryo SST anomalies can grow by a positive-feedback mechanism. We quote in full his explanation of the essential interaction mechanism:

In the presence of small oceanic thermal anomalies which tend to strengthen the latitudinal atmospheric temperature gradient, enhanced cyclone activity results. The rectified potential vorticity transport of the cyclone wave in regions of relatively strong mean temperature gradients tend to be balanced by vortex tube stretching in the atmosphere driven by heat exchange with the ocean. In an attempt to balance this heat exchange, meridional advection of sea-surface temperature by the wind-driven circulation is required. This in turn requires a change in the large-scale atmospheric thermal wind. The spatial phase in latitude of this change is such as to intensify the horizontal temperature gradient in which the cyclone wave is embedded, which leads to a further intensification
of the cyclone wave amplitude and sea-surface temperature anomaly. This constitutes a positive feedback in the coupled air–sea system.

These findings support the suggestion of Favorite and McClain (1973) that there is a slow instability in the ocean–atmosphere system which seemingly favours the enhancement of embryo SST anomalies. The anomalies may be initiated, Favorite and McClain thought, in western parts of oceans and grow by interaction with the atmosphere as they are carried along by the mean flow of the oceanic gyre containing them. Bretherton (1975) found it 'plausible that much of [the] eddy energy originates in the baroclinically unstable regions of the Gulf Stream and Kuroshio'.

It is clear from Bretherton's review paper that there have been some exciting oceanographic discoveries in recent years. What he called 'the classical picture' of large-scale oceanic circulations needs considerable revision. For example, Worthington (1977) has shown that there are *two* large-scale gyres in the North Atlantic Ocean, one entirely within the Sargasso Sea and the other south-east of Newfoundland (see Fig. 6.5).

To Bretherton, however, 'a major discovery of the last few years is that the subtropical oceans are full of quasi-two-dimensional turbulence, which in mid-ocean dominates the large-scale mean circulation and has an unknown influence upon it'. The Mid-Ocean Dynamics Experiment (MODE), a project concerning the British Institute of Oceanographic Sciences and several American organizations, was largely devoted to a study of this (meso-scale) turbulence; The observational phase of MODE took place in an area some 500 km across, south-west of Bermuda, during a period of 4 months in 1973. Further information about mid-ocean eddies has been derived by Gill (1975). Such information is, of course, essential if the eddies are to be modelled realistically.

Numerical models of oceanic circulations are, as WMO (1976) pointed out, generally less well developed than those of atmospheric behaviour. Nevertheless, some impressive advances have been made by dynamical oceanographers in the last decade or so. Those made at the GFDL are particularly impressive, as the papers of Kirk Bryan and his team show (see, for example: Bryan, 1969a; Bryan *et al.*, 1975). In the United Kingdom, a group in the Department of Applied Mathematics and Theoretical Physics of the University of Cambridge has consistently been among the leaders in the field of dynamical oceanography (see, for example Turner & Kraus, 1967; Gill *et al.*, 1974).

To further single out models for comment is an invidious task, and we apologise to those groups and individuals who may feel already that our summary in this chapter is unbalanced. For details of the many contributions which have been made to the problem of formulating mathematically the ocean—atmosphere system we suggest that reference be made to the bibliographies provided in works cited in this book. We may add to these a review paper by Gates (1975), on numerical modelling of climatic change, and the Proceedings of an international symposium (National Academy of Sciences, 1975) held at Durham, New Hampshire, in October 1972, the subject for discussion being numerical modelling of oceanic behaviour.

Towards the end of this symposium there took place a discussion entitled 'Where do we go from here?'. Hopefully, we provide in our discussion of projects and models in this chapter a glimpse of the course charted by distinguished meteorologists and oceanographers for research endeavours into the ocean—atmosphere system during the years up to about 1985.

Thirty years ago Sverdrup (1945) wrote:

It is not yet possible to deal with the system atmosphere—ocean as one unit, but it is obvious that, in treating separately the circulation of the atmosphere, a thorough consideration of the interaction between the atmosphere and the oceans is necessary.

These words have evidently been heeded, for, in recent years, as we stress in this book, much research has been directed towards the problems of air—sea interaction and ocean—atmosphere coupling, and some notable progress has been made by those seeking to model the unity which is the ocean—atmosphere system. Nevertheless, it is most unlikely that the system will be totally understood or completely modelled a decade hence, or maybe even half a century from now. The future of marine science is, therefore, exciting and challenging.

Notes

1. For a detailed assessment of the objectives and achievements of the Polar Year programmes and an outline of the evolution of the IGY concept, Gerson's (1958) article is recommended.
2. Etymologically, 'oceanology' means 'the study of oceanic phenomena' or 'oceanic science', but it is commonly taken to mean 'ocean technology'.
3. Conceptually, the World Weather Watch (WWW), according to WMO (1966), is: 'a world-wide meteorological system, composed of the co-ordinated national facilities and services provided by individual Members and supplemented by international organizations. Its primary purpose is to ensure that all Members obtain the meteorological information they require both for operational work and for research'. The elements of the WWW are (see WMO Secretariat, 1966): a global observing system, a global data-processing system, a global telecommunications system, a research programme, and a programme of education and training. The aims and aspirations of the WWW have been discussed by Bugaev (1973); plans and implementation programmes for the WWW have been given by the WMO Secretariat (1967, 1971); and technical details of the various elements of the WWW are to be found in a series of World Weather Watch Planning Reports, published by WMO.

   An Integrated Global Ocean Station System (IGOSS) is also being developed, to facilitate the international exchange of oceanographic data and information (see WMO Secretariat, 1973).

Appendix   The temperature–salinity diagram

Observations reveal that oceanic water masses possess characteristic associations of temperature and salinity, both areally and in the vertical. Water types can therefore be readily distinguished one from another by simple comparison of observed values of temperature and salinity.

Further, the density of a sample of water is determined principally by temperature and salinity, so that mutual relationships between temperature, salinity and density can be displayed graphically. Since variations of density with temperature and salinity are rather small, however, it is customary, for convenience, to make reference on a temperature–salinity (or T–S) diagram, not to water density (ρ_w) but to a parameter σ, defined by

$$\sigma = 10^3 (\rho_w - 1)$$

For most oceanographic purposes the small effect of pressure on water density can be disregarded, and the lines of σ shown on a T–S diagram are usually those of σ_t, the values of σ corresponding to atmospheric pressure.

T–S diagrams are used by oceanographers when studying water masses in much the same way as tephigrams or any other kinds of aerological diagram are used by meteorologists when studying atmospheric properties. More specifically, they are used for water-mass recognition, for ascertaining

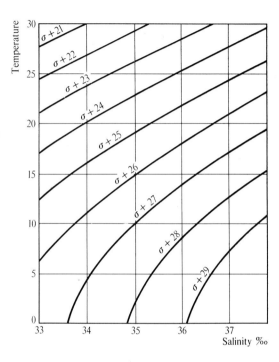

Appendix Diagram   Temperature–salinity diagram. (Temperature shown in °C)

the stability of water stratifications and for investigating mixing in the oceans (see Mamayev, 1975).

140

References

Adem, J. (1973) 'Ocean effects on weather and climate', *Geofisica Internacional*, **13**, 1–73.

Alaka, M. A. (1968) 'Climatology of Atlantic tropical storms and hurricanes', *ESSA Technical Report WB–6*, US Department of Commerce/Environmental Science Services Administration, 18 pp.

Aleem, A. A. (1967) 'Concepts of currents, tides and winds among medieval Arab geographers in the Indian Ocean', *Deep-Sea Res.*, **14**, 459–63.

Amos, A. F., Langseth, M. G. and Markl, R. G. (1972) 'Visible oceanic saline fronts', in *Studies in Physical Oceanography* (Wüst 80th birthday tribute, Ed. A. L. Gordon), Gordon & Breach, Vol. 1, 49–62.

Arx, W. S. von (1962) *An Introduction to Physical Oceanography,* Addison-Wesley, 422 pp.

Augstein, E., Schmidt, H. and Ostapoff, F. (1974) 'The vertical structure of the atmospheric planetary boundary layer in undistributed trade winds over the Atlantic Ocean', *Boundary-Layer Met.*, **6**, 129–50.

Badgley, F. I. (1966) 'Heat budget at the surface of the Arctic Ocean', *Proc. Symp. Arctic Heat Budget and Atmos. Circ.* (Ed. J. O. Fletcher), The Rand Corporation, Santa Monica, Calif., 267–77.

Badgley, P. C., Miloy, L. and Childs, L. (1969) *Oceans from Space,* Gulf Publ. Co. (Houston), 234 pp.

Bailey, H. S. (1953) 'The voyage of the *Challenger*', in *Oceanography: readings from Scientific American* (Ed. J. R. Moore), Freeman & Co. (1971), 20–4.

Baker, D. J. (1974) 'Heat balance of the Polar Oceans', in *US Contributions to the Polar Experiment (POLEX), Part 1,* National Academy of Sciences, Washington, DC, 70–9.

Ball, F. K. (1954) 'Sea-surface temperatures', *Australian J. Phys.*, **7**, 649–52.

Banke, E. G. and Smith, S. D. (1973) 'Wind stress on Arctic sea ice', *J. Geophys. Res.*, **78** (33), 7871–83.

Banke, E. G. and Smith, S. D. (1975) 'Measurement of form drag on ice ridges', *AIDJEX Bull.*, No. 28, Univ. Washington, Seattle, 21–8.

Bannister, R. C. (1948) 'A remarkable cold spell at Hong Kong', *Weather*, **3**, 344.

Barber, N. F. (1969) *Water Waves,* Wykeham Publications (London), 142 pp.

Barber, N. F. and Ursell, F. (1948) 'The generation and propagation of ocean waves and swell, I Wave periods and velocities', *Phil. Trans. R. Soc.* (A), **240** (824), 527–60.

Barkley, R. A. (1970) 'The Kuroshio Current', in *Oceanography: contemporary readings in ocean sciences* (Ed. R. G. Pirie), Oxford University Press (1973), 108–18.

Barnes, F. A. and King, C. A. M. (1953) 'The Lincolnshire coastline and the 1953 storm flood', *Geog.*, **38**, 141–60.

Barnett, T. P. (1973) 'The North Pacific Experiment: a study of large-scale ocean and atmosphere fluctuations in the Pacific', in *WMO Mar. Sci. Affairs Rep. No. 7* (Vol. 2), World Meteorological Organization (WMO – No. 350), 333–44.

Barrett, E. C. (1973) 'Daily and monthly rainfall estimates from weather satellite data', *Mon. Wea. Rev.*, **101**, 215–22.

Barrett, E. C. (1974) *Climatology from Satellites,* Methuen, 418 pp.

Barry, R. G. (1975) 'Climate models in palaeoclimatic reconstruction', *Palaeogeog. Palaeoclimat. Palaeoecol.*, **17**, 123–37.

Bascom, W. (1974) 'The disposal of waste in the ocean', *Sci. Amer.*, **231** (2), 16–25.

Bates, D. R. (1964) *The Planet Earth* (second edition), Pergamon, 370 pp.

Bathen, K. H. (1971) 'Heat storage and advection in the North Pacific Ocean', *J. Geophys. Res.*, **76** (3), 676–87.

Bénard, H. (1900) 'Les tourbillons cellulaires dans une nappe liquide', *Rev. Gén. Sci. Pures Appl.*, **11**, 1261–71, 1309–28.

Benton, G. S. and 7 co-authors (1963) 'Interaction between the atmosphere and the oceans', *Bull. Amer. Met. Soc.*, **44**, 4–17.

Bergsten, F. (1936) 'A contribution to the knowledge of the influence of the Gulf Stream on the winter temperature of northern Europe', *Medd. Met.-Hydr. Anst. Stockholm, Ser. Uppsats,* No. 10 (see also *Geograf. Ann.,* **3–4**, Stockholm).

Berry, F. A., Bollay, E. and Beers, N. R., eds. (1945) *Handbook of Meteorology,* McGraw-Hill, 1068 pp.

Bertalanffy, L. von (1956) 'General systems theory', in *General Systems Yearbook,* **1**, 1–10.

Biesel, F. (1952) 'Study of wave propagation in water of gradually varying depth', in *Gravity Waves,* US Nat. Bur. Stand. Circ. 521, 243–53.

Bjerknes, J. (1964) 'Atlantic air–sea interaction', *Advances in Geophysics,* **10**, 1–82.

Bjerknes, J. (1965) 'Atmosphere–ocean interaction during the "Little Ice Age" (seventeenth to nineteenth centuries, AD)', *WMO Tech. Note 66,* World Meteorological Organization, 77–88.

Bjerknes, J. (1966) 'A possible response of the atmospheric Hadley circulation to equatorial anomalies of ocean temperature', *Tellus,* **18**, 820–9.

Bjerknes, J. (1969) 'Atmospheric teleconnections from the equatorial Pacific', *Mon. Wea. Rev.,* **97**, 163–72.

Bjerknes, V. (1916) 'Über thermodynamische Maschinen,

die unter der Mitwirkung der Schwerkraft arbeiten', *Abb. Sächs. Akad. Wiss.*, 35 (1).

Black, P. G. and Mallinger, W. D. (1973) 'Use of satellite infrared imagery, airborne expendable bathythermographs and airborne precision infrared thermometers to infer the mutual interaction of hurricanes and the upper mixed layer of the ocean', in *WMO Mar. Sci. Affairs Rep.*, No. 7 (Vol. 2), World Meteorological Organization (WMO – No. 350), 290–314.

Blanchard, D. C. (1963) 'The electrification of the atmosphere by particles from bubbles in the sea', *Progress in Oceanography*, 1, Pergamon, 73–202.

Blanchard, D. C. and Woodcock, A. H. (1957) 'Bubble formation and modification in the sea and its meteorological significance', *Tellus*, 9, 145–58.

Blumer, M. (1971) 'Scientific aspects of the oil spill problem', in *Oceanography: contemporary readings in ocean sciences* (Ed. R. G. Pirie), Oxford University Press (1973), 453–64.

Bolin, B. R. (1970) 'Progress on the planning and implementation of the Global Atmospheric Research Programme', in *The Global Circulation of the Atmosphere* (Ed. G. A. Corby), Royal Meteorological Society (London), 235–55.

Bowden, K. F. (1962) 'Turbulence', in *The Sea*, Vol. 1 (Ed. M. N. Hill), Interscience Publ. (Wiley), 802–25.

Bowen, I. S. (1926) 'The ratio of heat losses by conduction and by evaporation from any water surface', *Phys. Rev.*, 27, 779–87.

Bretherton, F. P. (1975) 'Recent developments in dynamical oceanography', *Q. J. R. Met. Soc.*, 101, 705–21.

Bretschneider, C. L. (1952) 'Revised wave forecasting relationships', *Proc. IInd Conf. on Coastal Eng.*, Ch. 1, 1–5.

Bretschneider, C. L. (1957) Review of *Practical Methods for observing and forecasting ocean waves by means of wave spectra and statistics*, US Navy Dept. Hydrogr. Off. Pub. No. 603, in *Trans. Amer. Geophys. Union*, 264–6.

Bretschneider, C. L. (1959) 'Revisions in wave forecasting: deep and shallow water', *Proc. VIth Conf. on Coastal Eng.*, Ch. 3, 30–67.

Bretschneider, C. L. (1966) 'Wave generation by wind, deep and shallow water', in *Estuary and Coastline Hydrodynamics* (Ed. A. T. Ippen), McGraw-Hill, 133–96.

Bretschneider, C. L. (1970) 'Forecasting relations for wave generation', *Look Lab. Quarterly*, 1 (3), 31–4.

Brooks, C. E. P. and Hunt, T. M. (1930) 'The zonal distribution of rainfall over the earth', *Mem. R. Met. Soc.*, 3 (28), 139–58.

Brown, R. (1828) 'A brief account of microscopical observations made in the months of June, July, and August, 1827, on the particles contained in the pollen of plants; and on the general existence of active molecules in organic and inorganic bodies', *Phil. Mag.* (New Series), 4 (21), 161–73.

Bruce, J. G. (1974) 'Some details of upwelling off the Somali and Arabian coasts', *J. Mar. Res.*, 32, 419–23.

Brummage, K. G. (1973) 'What is marine pollution?', *Proc. Symp. Mar. Poll.*, R. Inst. Nav. Arch., 1–9.

Brummer, B., Augstein, E. and Riehl, H. (1974) 'On the low-level wind structure in the Atlantic trade', *Q. J. R. Met. Soc.*, 100, 109–21.

Bryan, K. (1969a) 'A numerical method for the study of the circulation of the World Ocean', *J. Computat. Phys.*, 4, 347–76.

Bryan, K. (1969b) 'Climate and the ocean circulation: III The ocean model', *Mon. Wea. Rev.*, 97, 806–27 (see also Manabe, 1969a, b).

Bryan, K. and Cox, M. D. (1968) 'A nonlinear model of an ocean driven by wind and differential heating', *J. Atmos. Sci.*, 25, 945–67, 968–78.

Bryan, K., Manabe, S. and Pacanowski, R. C. (1975) 'A global ocean–atmosphere climate model. Part II: the oceanic circulation', *J. Phys. Oceanogr.*, 5, 30–46.

Bryan, K. and Webster, J. (1960) 'Poleward heat transport of the North Atlantic and Pacific Oceans: A comparison of the estimates of Sverdrup and Budyko', unpublished MS on file at the Woods Hole Oceanogr. Inst., 7 pp (this reference taken from Malkus, 1962).

Bryant, G. W. and Browning, K. A. (1975) 'Multi-level measurements of turbulence over the sea during the passage of a frontal zone', *Q. J. R. Met. Soc.*, 101, 35–54.

Budyko, M. I. (1956) 'Teplovoǐ balans zemnoǐ poverkhnosti', *Gidrometeorologicheskoe izdatel'stvo*, Leningrad, 255 pp (US Weather Bureau translation: 'Heat balance of the earth's surface', 1958, PB131692, 259 pp).

Budyko, M. I. (1962) 'Polar ice and climate', *Bull. Acad. Sci. USSR*, Geog. Series No. 6, Moscow, 3–10.

Budyko, M. I., ed. (1963) *Atlas teplovogo balansa zemnogo shara*, Akad. Nauk SSSR, Moscow.

Budyko, M. I. (1966) 'Polar ice and climate', *Proc. Symp. Arctic Heat Budget and Atmos. Circ.* (Ed. J. O. Fletcher), The Rand Corporation, Santa Monica, California, 5–21.

Budyko, M. I. (1974) *Climate and Life* (English edn, Ed. D. H. Miller), Academic Press, 508 pp.

Budyko, M. I., Yefimova, N. A., Aubenok, L. I. and Strokina, L. A. (1962) 'The heat balance of the surface of the earth', *Sov. Geog.*, 3, 3–16.

Bugaev, V. A. (1973) 'The origin of the World Weather Watch and its future prospects', in *Tenth Anniversary of the World Weather Watch*, World Meteorological Organization (WMO – No. 342), 1–10.

Bullock, B. R. and Johnson, D. R. (1972) 'The generation of available potential energy by sensible heating in Southern Ocean cyclones', *Q. J. R. Met. Soc.*, 98, 495–518.

Bunker, A. F. (1960) 'Heat and water-vapor fluxes in air flowing southward over the western North Atlantic Ocean', *J. Met.*, 17, 52–63.

Bunker, A. F. (1975) 'Energy exchange at the surface of the western North Atlantic Ocean', *Woods Hole Oceanogr. Inst. Tech. Rep.*, 75 (3), 107 pp.

Burt, W. V., Cummings, T. and Paulson, C. A. (1974) 'The mesoscale wind field over the ocean', *J. Geophys. Res.*, 79 (36), 5625–32.

Busch, N. E. (1973) 'The surface boundary layer', *Boundary-Layer Met.*, 4, 213–40.

Campbell, W. J., Gloersen, P., Nordberg, W. and Wilheit, T. T. (1974) 'Dynamics and morphology of Beaufort Sea ice determined from satellites, aircraft, and drifting stations', *AIDJEX Bull.*, No. 25, University of Washington, Seattle, 1–28.

Carlson, T. N. (1969a) 'Synoptic histories of three African disturbances that developed into Atlantic hurricanes', *Mon. Wea. Rev.*, 97, 256–76.

Carlson, T. N. (1969b) 'Some remarks on African disturbances and their progress over the tropical Atlantic', *Mon. Wea. Rev.*, **97**, 716–26.

Carpenter, W. B. (1870) 'The Gulf Stream', letter to *Nature*, **2**, 334–5.

Carruthers, J. N. (1941) 'Some inter-relationships of meteorology and oceanography', *Q. J. R. Met. Soc.*, **67**, 207–46.

Cartwright, D. E. (1958) 'The scientific study of ship motions', *Sci. Prog.*, **46**, 83–91.

Cartwright, D. E. (1961) 'The ocean wave spectrum', *Sci. Prog.*, **49**, 681–93.

Caviedes, C. (1975) 'El Niño 1972: its climatic, ecological, human, and economic implications', *Geog. Rev.*, **65**, 493–509.

Champion, F. C. and Davy, N. (1959) *Properties of Matter* (third edition), Blackie & Son, 334 pp.

Chapman, S. (1964) 'The International Geophysical Year', in *The Planet Earth* (Ed. D. R. Bates, q.v.), 1–13.

Charney, J. G. (1955) 'The generation of oceanic currents by wind', *J. Mar. Res.*, **14**, 477–98.

Charnock, H. (1967) 'Flux-gradient relations near the ground in unstable conditions', *Q. J. R. Met. Soc.*, **93**, 97–100.

Cherry-Garrard, A. (1922) *The Worst Journey in the World*, published in Penguin Books (1970), 652 pp.

Chorley, R. J. and Kennedy, B. A. (1971) *Physical geography: a systems approach*, Prentice-Hall, 370 pp.

Clarke, G. L. (1967) 'Light in the sea', in *Oceanography: contemporary readings in ocean sciences* (Ed. R. G. Pirie), Oxford University Press (1973), 218–21.

Collins, C. A., Wilson, W. S., Carlmark, J., Merrell, W. J. and Barnett, T. P. (1973) 'The North Pacific Experiment probes weather phenomenon', *MTS Jour.*, **7** (1), 10–13.

Colon, J. A. (1964) 'On interactions between the Southwest Monsoon Current and the sea surface over the Arabian Sea', *Indian J. Met. Geophys.*, **15**, 183–200.

Coriolis, G. G. de (1835) 'Mémoire sur les équations du mouvement relatif des systèmes de corps', *J. Ec. Roy. Polyt.*, **15**, 142–54.

Corkan, R. H. (1948) *Storm surges in the North Sea*, US Hydrogr. Off., Misc. 15072, Vol. 1, 174 pp, Vol. 2, 166 pp.

Corkan, R. H. (1950) 'The levels in the North Sea associated with the storm disturbance of 8 January 1949', *Phil. Trans. R. Soc.* (A), **242** (853), 493–525.

Couper, A. D. (1974) Introduction to *The Celtic Sea: meteorological and oceanographic conditions*, occas. monogr. from Dept. Marit. Studs., Univ. Wales Inst. Sci. Tech., 1–6.

Cox, C. S. (1974) 'Refraction and reflection of light at the sea surface', in *Optical Aspects of Oceanography* (Eds. N. G. Jerlov and E. Steeman Nielsen, q.v.), 51–75.

Cox, C. S. and Munk, W. H. (1954) 'Statistics of the sea surface derived from sun glitter', *J. Mar. Res.*, **13**, 198–227.

Cox, M. D. (1970) 'A mathematical model of the Indian Ocean', *Deep-Sea Res.*, **17**, 47–75.

Craddock, J. M. (1951) 'The warming of Arctic air masses over the eastern North Atlantic', *Q. J. R. Met. Soc.*, **77**, 355–64.

Craddock, J. M. and Ward, R. (1962) 'Some statistical relationships between the temperature anomalies in neighbouring months in Europe and western Siberia',
Met. Off. (London) Sci. Pap., No. 12 (M.O. 718), 31 pp.

Criminale, W. O. and Spooner, G. F. (1975) 'Fluctuations and structure within the oceanic boundary layer below the Arctic ice cover', *AIDJEX Bull.*, No. 30, University of Washington, Seattle, 29–54.

Crowe, P. R. (1950) 'The seasonal variation in the strength of the trades', *Trans. Inst. Brit. Geog.*, **16**, 25–47.

Currie, R. (1966) 'Some reflections on the International Indian Ocean Expedition', *Oceanogr. Mar. Biol. Ann. Rev.*, **4**, 69–78.

Dampier, W. (1699) 'A discourse of winds, breezes, storms, tides and currents', in *A Collection of Voyages* (Vol. 2), 7th edn (1729), London.

Darbyshire, J. (1952) 'The generation of waves by wind', *Proc. R. Soc.* (A), **215** (1122), 299–328.

Darbyshire, J. (1955) 'An investigation of storm waves in the North Atlantic Ocean', *Proc. R. Soc.* (A), **230** (1183), 560–9.

Darbyshire, J. (1956) 'An investigation into the generation of waves when the fetch of the wind is less than 100 miles', *Q. J. R. Met. Soc.*, **82**, 461–8.

Darbyshire, J. (1959) 'A further investigation of wind generated waves', *Deut. Hydrogr. Zeit.*, **12** (1), 1–13.

Darbyshire, J. (1963) 'The one-dimensional wave spectrum in the Atlantic Ocean and in coastal waters', in *Ocean Wave Spectra*, Prentice-Hall, 27–31.

Darbyshire, J. and Darbyshire, M. (1955) 'Determination of wind stress on the surface of Lough Neagh by measurement of tilt', *Q. J. R. Met. Soc.*, **81**, 333–9.

Darbyshire, M. and Draper, L. (1963) 'Forecasting wind-generated sea waves', *Engineering*, **195**, 482–4.

Das, P. K., Sinha, M. C. and Balasubramanyam, V. (1974) 'Storm surges in the Bay of Bengal', *Q. J. R. Met. Soc.*, **100**, 437–49.

Davidson, B. (1968) 'The Barbados Oceanographic and Meteorological Experiment', *Bull. Amer. Met. Soc.*, **49**, 928–34.

Day, G. J. (1976) 'Progress in the development of meteorological buoy systems and marine automatic weather stations', *Mar. Obs.*, **46**, 22–7.

Day, J. A. (1964) 'Production of droplets and salt nuclei by bursting of air-bubble films', *Q. J. R. Met. Soc.*, **90**, 72–8.

Deacon, E. L. (1962) 'Aerodynamic roughness of the sea', *J. Geophys. Res.*, **67** (8), 3167–72.

Deacon, E. L. (1969) 'Physical processes near the surface of the earth', in *World Survey of Climatology*, Vol. 2 (Ed. H. Flohn), Elsevier, 39–104.

Deacon, E. L. and Webb, E. K. (1962) 'Small-scale interactions', in *The Sea*, Vol. 1 (Ed. M. N. Hill), Interscience Publ. (Wiley), 43–87.

Deacon, M. (1971) *Scientists and the Sea 1650–1900: a study of marine science*, Academic Press, 445 pp.

De Leonibus, P. S. (1971) 'Momentum flux and wave spectra observations from an ocean tower', *J. Geophys. Res.*, **76** (27), 6506–27.

Dietrich, G. and Kalle, K. (1957) *Allgemeine Meereskunde: eine Einführung in die Ozeanographie*, Gebrüder Borntraeger (Berlin), 492 pp.

Dietz, R. S. and LaFond, E. C. (1950) 'Natural slicks on the ocean', *J. Mar. Res.*, **9**, 69–76.

Dines, J. S. (1929) 'Meteorological conditions associated with high tides in the Thames', *Met. Off. (London) Geophys. Mem.*, No. 47, 27–39.

Dines, W. H. (1917) 'The heat balance of the atmosphere', *Q. J. R. Met. Soc.*, **43**, 151–8.

Dinklage, L. E. (1874) Unpublished paper, quoted at length by Schott, G. (1891) in 'Die Meeresströmungen und Temperaturverhältnisse in den Ostasiatischen Gewässern', *Petermans Mitt.*, **37**, 209–19. Schott noted that Dinklage's paper was lodged in the archives of Deutsche Seewarte in Hamburg.

Dixit, B. and **Pounder, E. R.** (1975) 'The specific heat of saline ice', *J. Glaciol.*, **14**, 459–65.

Donn, W. L. and **Ewing, M.** (1968) 'The theory of an ice-free Arctic Ocean', *Met. Monogr.*, **8** (30), 100–5.

Doodson, A. T. (1929) 'Report on Thames floods', *Met. Off. (London) Geophys. Mem.*, No. 47, 3–26.

Draper, L. (1965) 'Wave spectra provide best basis for offshore rig design', *Oil and Gas Internat.*, **5** (6), 58–60.

Draper, L. (1967) 'The analysis and presentation of wave data – a plea for uniformity', *Proc. 10th Conf. Coastal Eng.* (Amer. Soc. Civ. Eng.), **1**, 1–11.

Draper, L. (1970a) 'Routine sea-wave measurement – a survey', *Underwater Sci. Tech. J.*, **2**, 81–6.

Draper, L. (1970b) 'Environmental conditions', *Proc. Symp. Offshore Drilling Rigs* (R. Inst. Nav. Arch.), 1–12.

Draper, L. (1974) 'Oceanographic conditions', in *The Celtic Sea: meteorological and oceanographic conditions,* occas. monogr. from Dept. Marit. Studs., Univ. Wales Inst. Sci. Tech., 87–121.

Draper, L. and **Tucker, M. J.** (1971) 'The determination of wave conditions for marine engineering', in *Dynamic Waves in Civil Engineering* (Ed. D. A. Howells *et al.*), Wiley-Interscience, 389–402.

Drozdov, O. A. and **Berlin, I. A.** (1953) 'Annual total precipitation', *Morskoi Atlas* (Vol. 2), sheet 48.

Düing, W. (1970) *The monsoon régime of the currents in the Indian Ocean,* Internat. Indian Oc. Exped. Oceanogr. Monogr. No. 1, East-West Center Press (Honolulu), 68 pp.

Dunbar, M. and **Wittmann, W.** (1963) 'Some features of ice movement in the Arctic basin', *Proc. Arctic Basin Symp.*, Arctic Inst. N. Amer., Washington, 90–108.

Dunn, G. E. and **Miller, B. I.** (1964) *Atlantic Hurricanes,* Louisiana State University Press, 377 pp.

Dury, J. M. (1970) 'The Beaufort scale of wind force', *WMO Mar. Sci. Affairs Rep. No. 3,* World Meteorological Organization, 22 pp.

Eady, E. T. (1964) 'The general circulation of the atmosphere and oceans', in *The Planet Earth* (Ed. D. R. Bates, q.v.), 141–63.

Eckart, C. (1953) 'The generation of wind waves on a water surface', *J. Appl. Phys.*, **24**, 1485–94.

Ehrlich, P. R. and **Ehrlich, A. H.** (1972) *Population, Resources, Environment* (second edn), W. H. Freeman & Co., San Francisco, 509 pp.

Ekman, V. W. (1902) 'Om jordrotationens inverkan på vindströmmar i hafvet', *Nyt. Mag. f. Naturvid.*, **20**, Kristiania, 20 pp.

Ekman, V. W. (1905) 'On the influence of the Earth's rotation on ocean currents', *Ark. f. Math., Astron. och Fysik,* **2** (11), 52 pp.

Ekman, V. W. (1923) 'Über Horizontalzirkulation bei winderzeugten Meeresströmungen', *Ark. f. Math., Astron. och Fysik,* **17** (26), 74 pp.

Ekman, V. W. (1932) 'Studien zur Dynamik der Meeresströmungen', *Gerl. Beitr. z. Geophys.*, **36**, 385–438.

Emig, M. (1967) 'Heat transport by ocean currents', *J. Geophys. Res.*, **72** (10), 2519–29.

Eriksson, E. (1959) 'The yearly circulation of chloride and sulfur in nature; meteorological, geochemical and pedological implications', *Tellus,* **11**, 375–403, and **12** (1960), 63–109.

Evans, S. H. (1968) 'Weather routeing of ships', *Weather,* **23**, 2–8.

Ewing, G. (1950) 'Slicks, surface films and internal waves', *J. Mar. Res.*, **9**, 161–87.

Faegri, K. (1950) 'On the value of palaeoclimatological evidence', *Cent. Proc. R. Met. Soc.*, 188–95.

Farmer, S. A. (1973) 'A note on the long term effects on the atmosphere of sea-surface temperature anomalies in the North Pacific Ocean', *Weather,* **28**, 102–5.

Favorite, F. and **McClain, D. R.** (1973) 'Coherence in trans-Pacific movements of positive and negative anomalies of sea-surface temperature 1953–1960', *Nature,* **244**, 139–43.

Fendell, F. E. (1974) 'Tropical cyclones', *Advances in Geophysics,* **17**, 1–100.

Ferrel, W. (1856) 'An essay on the winds and currents of the ocean', *Nashville J. Med. Surg.*, **11**, 277–301, 375–89.

Ferrel, W. (1859–60) 'The motions of fluids and solids relative to the earth's surface', *Math. Monthly:* **1**, 140–8, 210–16, 300–7, 366–73, 397–406; **2**, 85–97, 339–46, 374–82.

Ferrel, W. (1861) 'The motions of fluids and solids relative to the earth's surface', *Amer. J. Sci.*, **31**, 27–57.

Ferrel, W. (1889) *A popular treatise on the winds,* Wiley, 505 pp.

Findlay, A. G. (1869) 'On the Gulf Stream', *Proc. R. Geog. Soc.*, **13**, 102–12.

Finizio, C., Palmieri, S. and **Riccucci, A.** (1972) 'A numerical model of the Adriatic for the prediction of high tides at Venice', *Q. J. R. Met. Soc.*, **98**, 86–104.

Flather, R. A. and **Davies, A. M.** (1976) 'Note on a preliminary scheme for storm surge prediction using numerical models', *Q. J. R. Met. Soc.*, **102**, 123–32.

Fleagle, R. G. (1956) 'The temperature distribution near a cold surface', *J. Met.*, **13**, 160–5.

Fletcher, J. O. (1965) *The heat budget of the Arctic Basin and its relation to climate,* Rand Corporation Rep. R-444-PR, Santa Monica, Calif., 179 pp.

Flohn, H. (1953) 'Wilhelm Meinardus und die Revision unserer Vorstellung von der atmosphärischen Zirkulation', *Zeit. Met.*, **7**, 97–108.

Flohn, H. (1964) 'Investigations on the tropical easterly jet', *Bonn. Met. Abh.*, No. 4, 83 pp.

Flohn, H. (1972) 'Investigations of equatorial upwelling and its climatic rôle', *Studies in Physical Oceanography* (Wüst 80th birthday tribute, Ed. A. L. Gordon), publ. by Gordon & Breach, Vol. 1, 93–102.

Flohn, H. (1973) 'Remarks on climatic intransitivity and the 1972 Pacific anomaly', *Atmosphere,* **11**, 134–40.

Flohn, H. (1975) 'Climatic teleconnections with the equatorial Pacific and the rôle of ocean/atmosphere coupling', *Atmosphere,* **13**, 96–109.

Forbes, R. J. and **Dijksterhuis, E. J.** (1963) *A History of Science and Technology* (Vol. 1), Pelican Books, 294 pp.

Francis, J. R. D. (1954) 'Wind stress on a water surface', *Q. J. R. Met. Soc.*, **80**, 438–43.

Franklin, B., Brownrigg, W. and Mr Farish (1774) 'Of the stilling of waves by means of oil', *Phil. Trans. R. Soc.,* **64,** 445–60.

Fuglister, F. C. (1955) 'Alternative analyses of current surveys', *Deep-Sea Res.,* **2,** 213–29.

Fuglister, F. C. (1963) 'Gulf Stream '60', in *Progress in Oceanography,* Vol. 1 (Ed. M. Sears), Pergamon, 265–373.

Fuglister, F. C. and Worthington, L. V. (1951) 'Some results of a multiple ship survey of the Gulf Stream', *Tellus,* **3,** 1–14.

Gagnon, R. M. (1964) *Types of winter energy budgets over the Norwegian Sea,* Arctic Met. Res. Group McGill Univ. Publ. in met. No. 64, 56 pp.

Gall, R. L. and Johnson, D. R. (1971) 'The generation of available potential energy by sensible heating: a case study', *Tellus,* **23,** 465–82.

Garbett, L. G. (1926) 'Admiral Sir Francis Beaufort and the Beaufort Scales of Wind and Weather', *Q. J. R. Met. Soc.,* **52,** 161–72.

García, R. V. (1968) 'First session of the Joint GARP Organizing Committee', *WMO Bull.,* **17,** 126–30.

Garstang, M. (1967) 'Sensible and latent heat exchange in low latitude synoptic scale systems', *Tellus,* **19,** 492–508.

Garstang, M. (1972) 'A review of hurricane and tropical meteorology', *Bull. Amer. Met. Soc.,* **53,** 612–30.

Garstang, M., LaSeur, N. E., Warsh, K. L., Hadlock, R. and Petersen, J. R. (1970) 'Atmospheric–oceanic observations in the tropics', in *Oceanography: contemporary readings in ocean sciences* (Ed. R. G. Pirie), Oxford University Press (1973), 177–95.

Gates, D. M. (1962) *Energy exchanges in the biosphere,* Harper & Row.

Gates, W. L. (1975) *Numerical modelling of climatic change: a review of problems and prospects,* The Rand Corporation, Santa Monica, Calif. Paper P-5471, 24 pp.

Gerson, N. C. (1958) 'From Polar Years to IGY', *Advances in Geophysics,* **5,** 1–52.

Gill, A. E. (1973) 'Circulation and bottom water production in the Weddell Sea', *Deep-Sea Res.,* **20,** 111–40.

Gill, A. E. (1975) 'Evidence for mid-ocean eddies in weather ship records', *Deep-Sea Res.,* **22,** 647–52.

Gill, A. E., Green, J. S. A. and Simmons, A. J. (1974) 'Energy partition in the large-scale ocean circulation and the production of mid-ocean eddies', *Deep-Sea Res.,* **21,** 499–528.

Gloersen, P. and Salomonson, V. V. (1975) 'Satellites – new global observing techniques for ice and snow', *J. Glaciol.,* **15,** 373–89.

Golden, J. H. (1968) 'Waterspouts at Lower Matecumbe Key, Florida, 2 September 1967', *Weather,* **23,** 103–14.

Golden, J. H. (1973) 'Some statistical aspects of waterspout formation', *Weatherwise,* **26,** 108–17.

Gordienko, P. A. (1961) 'The Arctic Ocean', in *Oceanography: readings from Scientific American* (Ed. J. R. Moore), Freeman & Co. (1971), 92–104.

Gordon, A. H. (1951) 'Waterspouts', *Mar. Obs.,* **21,** 47–60, 87–93.

Gordon, A. H. (1952) 'Relation of atmospheric humidity at low levels over the sea to wind force and the difference in temperature between air and sea', *Met. Mag.,* **81,** 289–95.

Gordon, A. L. and Tchernia, P. (1972) 'Waters of the continental margin off Adélie Coast, Antarctica', *Antarctic Res. Ser.* (Amer. Geophys. Un.), **19,** 59–69.

Gordon, A. L. (1975) 'General ocean circulation', in *Numerical Models of Ocean Circulation* (National Academy of Sciences, 1975, q.v.), 39–53.

Gow, A. J. (1965) 'The ice sheet', in *Antarctica* (Ed. T. Hatherton), Methuen, 221–58.

Gray, W. M. (1967) 'The mutual variation of wind, shear, and baroclinicity in the cumulus convective atmosphere of the hurricane', *Mon. Wea. Rev.,* **95,** 55–73.

Gray, W. M. (1968) 'Global view of the origin of tropical disturbances and storms', *Mon. Wea. Rev.,* **96,** 669–700.

Gray, W. M. (1975) *Tropical cyclone genesis,* Colorado State Univ. Atm. Sci. Pap. No. 234, 121 pp.

Green, J. S. A., Ludlam, F. H. and McIlveen, J. F. R. (1966) 'Isentropic relative-flow analysis and the parcel theory', *Q. J. R. Met. Soc.,* **92,** 210–19.

Grieve, H. (1959) *The Great Tide,* Essex County Council (Chelmsford), 883 pp.

Groen, P. (1967) *The Waters of the Sea,* Van Nostrand, 328 pp.

Gunther, E. R. (1936) 'A report on oceanographical investigations in the Peru Coastal Current', *Discovery Rep.* (Cambridge University Press), **13,** 107–276.

Hadley, G. (1735) 'Concerning the cause of the general trade winds', *Phil. Trans. R. Soc.,* **39,** 58–62.

Hales, S. (1758) *A Treatise on Ventilators,* London.

Hall, A. D. and Fagen, R. E. (1968) 'Definition of System', in *Modern Systems Research for the Behavioral Scientist* (Ed. W. Buckley), Aldine Publ. Co. (Chicago), 81–92.

Halley, E. (1686) 'An historical account of the trade winds and monsoons, observable in the seas between and near the tropicks; with an attempt to assign the phisical cause of the said winds', *Phil. Trans. R. Soc.,* **16,** 153–68.

Hantel, M. (1970) 'Monthly charts of surface wind stress curl over the Indian Ocean', *Mon. Wea. Rev.,* **98,** 765–73.

Hantel, M. (1971) 'Surface wind vergence over the tropical Indian Ocean', *J. Appl. Met.,* **10,** 875–81.

Hantel, M. (1972) 'Wind stress curl – the forcing function for oceanic motions', in *Studies in Physical Oceanography* (Wüst 80th birthday tribute, Ed. A. L. Gordon), publ. by Gordon & Breach, Vol. 1, 121–36.

Hare, F. K. (1966) 'The concept of climate', *Geog.,* **51,** 99–110.

Harrold, T. W. and Browning, K. A. (1969) 'The polar low as a baroclinic disturbance', *Q. J. R. Met. Soc.,* **95,** 710–23.

Hasse, L. (1963) 'On the cooling of the sea surface by evaporation and heat exchange', *Tellus,* **15,** 363–6.

Hasselmann, K. (1974) 'On the spectral dissipation of ocean waves due to white capping', *Boundary-Layer Met.,* **6,** 107–27.

Hasselmann, K. and 15 co-authors (1973) *Measurements of wind-wave growth and swell decay during the Joint North Sea Wave Project (JONSWAP),* Deutsches Hydrogr. Inst. (Hamburg), Ergänzungsheft zur Deutsches Hydrogr. Zeit., Reihe A, Nr. 12, 95 pp.

Hay, R. F. M. (1953) 'Frost smoke and unusually low air temperature at Ocean Weather Station India', *Mar.Obs.,* **23,** 218–25.

Heap, J. A. (1965) 'Antarctic Pack Ice', in *Antarctica* (Ed. T. Hatherton), Methuen, 187–96.

Heaps, N. S. (1967) 'Storm surges', *Oceanogr. Mar. Biol. Ann. Rev.*, **5**, 11–47.

Heaps, N. S. (1969) 'A two-dimensional sea model', *Phil. Trans. R. Soc.* (A), **265**, 93–137.

Helland-Hansen, B. (1916) 'Nogen hydrografiske metoder', *Forh. Skand. naturf. Møte* **16**, 357–9.

Helland-Hansen, B. and Nansen, F. (1920) 'Temperature variations in the North Atlantic Ocean and in the atmosphere: introductory studies on the causes of climatological variations', *Misc. Coll.*, **70**, Smithsonian Inst. Publ. 2537, 408 pp.

Herdman, W. A. (1923) *Founders of Oceanography and their Work: an Introduction to the Science of the Sea*, Edward Arnold, 340 pp.

Hidy, G. M. (1972) 'A view of recent air–sea interaction research', *Bull. Amer. Met. Soc.*, **53**, 1083–1102.

Hobbs, P. V. (1971) 'Simultaneous airborne measurements of cloud condensation nuclei and sodium-containing particles over the ocean', *Q. J. R. Met. Soc.*, **97**, 263–71.

Holland, J. Z. (1972) 'Comparative evaluations of some BOMEX measurements of sea-surface evaporation energy flux and stress', *J. Phys. Oceanogr.*, **2**, 476–86.

Houghton, D. D., Kutzbach, J. E., McClintock, M. and Suchman, D. (1974) 'Response of a general circulation model to a sea temperature perturbation', *J. Atmos. Sci.*, **31**, 857–68.

Houghton, D. M. (1969) 'Acapulco '68', *Weather*, **24**, 2–18.

Houghton, H. (1954) 'On the annual heat balance of the Northern Hemisphere', *J. Met.*, **11**, 1–9.

Houghton, J. T. and Taylor, F. W. (1973) 'Remote sounding from artificial satellites and space probes of the atmospheres of the earth and the planets', *Rep. Prog. Phys.*, **36**, 827–919.

Huang, J. C. K. (1975) 'Mid-latitude sea-surface temperature anomaly experiments with NCAR general circulation model', *Proc. Symp. Long-term Climatic Fluctuations* (Norwich), WMO – No. 421, 399–406.

Hubert, L. F. (1966) *Mesoscale cellular convection*, US Dept. Commerce Env. Sci. Serv. Admin. Met. Satellite Lab. Rep. No. 37, 68 pp.

Humboldt, A. von (1811) *Essai Politique sur le royaume de la Nouvelle-Espagne*, 5 vols., Paris.

Humboldt, A. von (1814) *Voyage aux régions équinoxiales du nouveau continent, fait en 1799–1804*, 3 vols., Paris.

Hunt, R. D. (1972) 'North Sea storm surges', *Mar. Obs.*, **42**, 115–24.

Ichiye, T. (1965) 'Diffusion experiments in coastal waters using dye techniques', *Proc. Sump. Diffusion in Oceans and Fresh Waters, 1964, Lamont Geol. Obsy., Palisades, New York*, 54–67.

Isaacs, J. D. (1969) 'The nature of oceanic life', in *Oceanography: readings from Scientific American* (Ed. J. R. Moore), Freeman & Co. (1971), 215–27.

Iselin, C. O'D. (1936) 'A case study of the circulation of the western North Atlantic', *Papers in phys. oceanogr. & met.* (Mass. Inst. Tech. and Woods Hole Oceanogr. Inst.), **4** (4), 101 pp.

Jacobs, S. S., Amos, A. F. and Bruchhausen, P. M. (1970) 'Ross Sea oceanography and Antarctic Bottom Water formation', *Deep-Sea Res.*, **17**, 935–62.

Jacobs, W. C. (1942) 'On the energy exchange between sea and atmosphere', *J. Mar. Res.*, **5**, 37–66.

Jacobs, W. C. (1951) 'The energy exchange between sea and atmosphere and some of its consequences', *Bull. Scripps Inst. Oceanogr.*, **6**, 27–122.

James, R. W. (1957) *Application of wave forecasts to marine navigation*, US Naval Oceanogr. Off. Pap. SP–1, 85 pp.

Jeffreys, H. (1925) 'On the formation of water waves by wind', *Proc. R. Soc.* (A), **107**, 189–206; **110**, 241–7.

Jerlov, N. G. and Steemann Nielsen, E. (1974) *Optical Aspects of Oceanography*, Academic Press, 494 pp.

Johnson, D. H. (1970) 'The rôle of the tropics in the global circulation', in *The Global Circulation of the Atmosphere* (Ed. G. A. Corby), Royal Meteorological Society (London), 113–36.

Johnson, D. R. (1970) 'The available potential energy of storms', *J. Atmos. Sci.*, **27**, 727–41.

Junge, C. E. (1963) *Air Chemistry and Radioactivity*, Academic Press, 382 pp.

Junge, C. E. (1972) 'Our knowledge of the physico-chemistry of aerosols in the undisturbed marine environment', *J. Geophys. Res.*, **77** (27), 5183–200.

Kamburova, P. L. and Ludlam, F. H. (1966) 'Rainfall evaporation in thunderstorm downdraughts', *Q. J. R. Met. Soc.*, **92**, 510–18.

Keers, J. F. (1966) 'The meteorological conditions leading to storm surges in the North Sea', *Met. Mag.*, **95**, 261–72.

Keers, J. F. (1968) 'An empirical investigation of interaction between storm surge and astronomical tide on the east coast of Great Britain', *Deutsches Hydrogr. Zeit.*, **21**, 118–25.

Kellogg, W. W. (1974) 'Meteorological observations in support of weather forecasting', in *US Contribution to the Polar Experiment (POLEX), Part 1*, National Academy of Sciences, Washington, DC, 26–49.

Kinsman, B. (1965) *Wind Waves: their generation and propagation on the ocean surface*, Prentice-Hall, 676 pp.

Kirwan, R. (1787) *An Estimate of the Temperature of Different Latitudes*, London.

Kitaigorodskii, S. A. (1970) *The physics of air–sea interaction*, Gidrometeorologicheskoe Izdatel'stvo, Leningrad; trans. from the Russian by A. Baruch, Israel Program for Sci. Trans (Jerusalem), 1973, 237 pp.

Koerner, R. M. (1971) 'Ice balance in the Arctic Ocean', *AIDJEX Bull.*, No. 6, University of Washington, Seattle, 11–26.

Koerner, R. M. (1973) 'The mass balance of the sea ice of the Arctic Ocean', *J. Glaciol.*, **12**, 173–85.

Kolesnikov, A. G. (1958) 'On the growth rate of sea-ice', *Arctic Sea Ice*, Proc. Maryland Conf., Nat. Acad. Sci. – Nat. Res. Coun. Publ. No. 598, 157–61.

Kondratyev, K. Ya. (1972) *Radiation processes in the atmosphere*, World Meteorological Organization (WMO – No. 309), 214 pp.

Kondratyev, K. Ya. and 4 co-authors (1970) 'Complex Energetics Experiment (CENEX)', *WMO Bull.*, **19**, 217–22.

Kondratyev, K. Ya. and 5 co-authors (1973) 'Some results of investigations under the programme of the Complex Atmospheric Energetics Experiment (1970–1972)', *WMO Bull.*, **22**, 7–13.

146

Kort, V. G. (1962) 'The Antarctic Ocean', in *Oceanography: readings from Scientific American* (Ed. J. R. Moore), Freeman & Co. (1971), 83–91.

Kraus, E. B. (1968) 'What we do not know about the sea-surface wind stress', *Bull. Amer. Met. Soc.,* **49**, 247–53.

Kraus, E. B. (1972) *Atmosphere–Ocean Interaction,* Clarendon Press (Oxford), 275 pp.

Kraus, E. B. and Morrison, R. E. (1966) 'Local interactions between the sea and the air at monthly and annual time scales', *Q. J. R. Met. Soc.,* **92**, 114–27.

Kuettner, J. P. (1959) 'The band structure of the atmosphere', *Tellus,* **11**, 267–94.

Kuettner, J. P. (1971) 'Cloud bands in the earth's atmosphere', *Tellus,* **23**, 404–25.

Kuettner, J. P. (1974) 'General description and central programme of GATE', *Bull. Amer. Met. Soc.,* **55**, 712–19.

Kuettner, J. P. and Parker, D. E. (1976) 'GATE: report on the field phase', *Bull. Amer. Met. Soc.,* **57**, 11–27.

Laevastu, T. (1965) 'Daily heat exchange in the North Pacific, its relations to weather and its oceanographic consequences', *Commentat. Phys.-Math.,* **31**, 5–53.

Laevastu, T. and Hela, I. (1970) *Fisheries Oceanography,* Fishing News (Books) Ltd. (London), 238 pp.

LaFond, E. C. (1959) 'Sea-surface features and internal waves in the sea', *Indian J. Met. Geophys.,* **10**, 415–19.

Lamb, H. (1932) *Hydrodynamics* (6th edn), Cambridge University Press (1975 reprint) 738 pp.

Lamb, H. H. (1972) *Climate: Present, Past and Future* (Vol. 1), Methuen, 613 pp.

Lamb, H. H. and Johnson, A. I. (1959) 'Climatic variation and observed changes in the general circulation', *Geograf. Ann.,* **41**, 94–134.

LaMer, V. K., ed. (1962) *Retardation of evaporation by monolayers: transport processes,* Academic Press, 277 pp.

Lane, F. W. (1968) *The Elements Rage* (Vol. 2), Sphere Books Ltd. (London), 156 pp.

Langmuir, I. (1938) 'Surface motion of water induced by wind', *Science,* **87**, (2250), 119–23.

Lee, A. and Ellett, D. (1967) 'On the water masses of the northwest Atlantic Ocean', *Deep-Sea Res.,* **14**, 183–90.

Leighly, J. (1968) 'M. F. Maury in his time', *Bull. Inst. Océanogr. Monaco,* No. spécial 2, 147–61.

Leipper, D. F. (1967) 'Observed ocean conditions and hurricane Hilda, 1964', *J. Atmos. Sci.,* **24**, 182–96.

Lemone, M. A. (1973) 'The structure and dynamics of horizontal roll vortices in the planetary boundary layer', *J. Atmos. Sci.,* **30**, 1077–91.

Lennon, G. W. (1963) 'The identification of weather conditions associated with the generation of major storm surges along the west coast of the British Isles', *Q. J. R. Met. Soc.,* **89**, 381–94.

Lenschow, D. H. (1972) 'The Air Mass Transformation Experiment (AMTEX)', *Bull. Amer. Met. Soc.,* **53**, 353–7.

Lenschow, D. H. (1974) 'The Air Mass Transformation Experiment (AMTEX): preliminary results from 1974 and plans for 1975', *Bull. Amer. Met. Soc.,* **55**, 1228–35.

Lewis, E. L. and Weeks, W. F. (1970) 'Sea ice: some polar contrasts', *Proc. Tokyo Symp. Antarctic Ice and Water Masses,* Cold Regions Res. & Eng. Lab. (Hanover, New Hampshire), 23–34.

Lighthill, M. J. (1969a) 'Unsteady wind-driven currents', *Q. J. R. Met. Soc.,* **95**, 675–88.

Lighthill, M. J. (1969b) 'Dynamic response of the Indian Ocean to onset of the Southwest Monsoon', *Phil. Trans. R. Soc.* (A), **265** (1159), 45–92.

London, J. (1957) *A study of the atmospheric heat balance,* N.Y. Univ., Dept. Met. Oceanogr. Final Rep., Project 131, Contract No. AF19(122)-165, 99 pp.

Longuet-Higgins, M. S. (1952) 'On the statistical distribution of the heights of sea waves', *J. Mar. Res.,* **11**, 245–66.

Longuet-Higgins, M. S. (1957) 'The statistical analysis of a random moving surface', *Phil. Trans. R. Soc.* (A), **249** (966), 321–87.

Longuet-Higgins, M. S. (1962) 'The directional spectrum of ocean waves, and processes of wave generation', *Proc. R. Soc.* (A), **265** (1322), 286–315.

Longuet-Higgins, M. S. (1965) 'Some dynamical aspects of ocean currents', *Q. J. R. Met. Soc.,* **91**, 425–51.

Longuet-Higgins, M. S. (1969) 'A nonlinear mechanism for the generation of sea waves', *Proc. R. Soc.* (A), **311** (1506), 371–89.

Lorenz, E. N. (1967) *The Nature and Theory of the General Circulation of the Atmosphere,* World Meteorological Organization, 161 pp.

Lorenz, E. N. (1968) 'Climatic determinism', *Met. Monogr.,* **8** (30), 1–3.

Lorenz, E. N. (1970) 'Climatic change as a mathematical problem', *J. Appl. Met.,* **9**, 325–9.

Ludlam, F. H. (1963) 'Severe local storms: a review', *Met. Monogr.,* **5** (27), 1–32.

Ludlam, F. H. (1966a) 'The cyclone problem: a history of models of the cyclonic storm', *Univ. London (Imperial Coll.) Professorial Inaugural Lect.,* 19–49.

Ludlam, F. H. (1966b) 'Cumulus and cumulonimbus convection', *Tellus,* **18**, 687–98.

Ludlam, F. H. and Scorer, R. S. (1957) *Cloud Study,* John Murray, 80 pp.

Lumb, F. E. (1961) 'Seasonal variation of the sea-surface temperature in coastal waters of the British Isles', *Met. Off. (London) Sci. Pap.,* No. 6 (M.O.685), 21 pp.

Lumley, J. L. and Panofsky, H. A. (1964) *The Structure of Atmospheric Turbulence,* Interscience Monogr. & Texts Phys. Astron. Vol. 12 (Wiley), 239 pp.

Lyall, I. T. (1972) 'The polar low over Britain', *Weather,* **27**, 378–90.

MJR (1975) 'Global Atmospheric Research Programme', *WMO Bull.,* **24**, 121–3.

MWS (1975) 'Global Atmospheric Research Programme: progress of the GARP Atlantic Tropical Experiment', *WMO Bull.,* **24**, 35–7.

Malkus, J. S. (1952) 'Recent advances in the study of convective clouds and their interaction with the environment', *Tellus,* **4**, 71–87.

Malkus, J. S. (1957) 'Trade cumulus cloud groups: some observations suggesting a mechanism of their origin', *Tellus,* **9**, 33–44.

Malkus, J. S. (1958) 'On the structure of the trade-wind moist layer', *Papers in phys. oceanogr. & met.* (Mass. Inst. Tech. and Woods Hole Oceanogr. Inst.), **13**, 47 pp.

Malkus, J. S. (1962) 'Large-scale interactions', in *The Sea,* Vol. 1 (Ed. M. N. Hill), Interscience Publ. (Wiley), 88–294.

Mallinger, W. D. and Mickelson, T. P. (1973) 'Experiments with monomolecular films on the surface of the open sea', *J. Phys. Oceanogr.*, **3**, 328–36.

Mallory, J. K. (1974) 'Abnormal waves on the south-east coast of South Africa', *Int. Hydrogr. Rev.*, **51**, 99–129.

Mamayev, O. I. (1975) *Temperature–Salinity Analysis of World Ocean Waters*, Elsevier Oceanogr. Ser., II (trans. from Russian by R. J. Burton), **11**, 374 pp.

Manabe, S. (1957) 'On the modification of air-mass over the Japan Sea when the outburst of cold air predominates', *J. Met. Soc. Japan* (Ser. 2), **35**, 311–26.

Manabe, S. (1969a) 'Climate and the ocean circulation: I. The atmospheric circulation and the hydrology of the earth's surface', *Mon. Wea. Rev.*, **97**, 739–74.

Manabe, S. (1969b) 'Climate and the ocean circulation: II. The atmospheric circulation and the effect of heat transfer by ocean currents', *Mon. Wea. Rev.*, **97**, 775–805.

Manabe, S. and Bryan, K. (1969) 'Climate calculations with a combined ocean–atmosphere model', *J. Atm. Sci.*, **26**, 786–9.

Manabe, S., Bryan, K. and Spelman, M. J. (1975) 'A global ocean–atmosphere climate model', *J. Phys. Oceanogr.*, **5**, 3–29.

Manabe, S., Hahn, D. G. and Holloway, J. L. (1974) 'The seasonal variation of the tropical circulation as simulated by a global model of the atmosphere', *J. Atmos. Sci.*, **31**, 43–83.

Manabe, S. and Holloway, J. L. (1975) 'The seasonal variation of the hydrologic cycle as simulated by a global model of the atmosphere', *J. Geophys. Res.*, **80** (12), 1617–49.

Manabe, S., Smagorinsky, J. and Strickler, R. F. (1965) 'Simulated climatology of a general circulation model with a hydrologic cycle', *Mon. Wea. Rev.*, **93**, 769–98.

Manier, G. and Moller, F. (1961) *Determination of the heat balance of the boundary layer over the sea,* Joh. Guttenberg Univ. Mainz, Final Rep. Contract No. AF 61(052)-315.

Mansfield, D. A. (1974) 'Polar lows: the development of baroclinic disturbances in cold air outbreaks', *Q. J. R. Met. Soc.*, **100**, 541–54.

Martin, D. W. and Sikdar, D. N. (1975) 'A case study of Atlantic cloud clusters: Part 1. Morphology and thermodynamic structure', *Mon. Wea. Rev.*, **103**, 691–708.

Mason, B. J. (1954) 'Bursting of air bubbles at the surface of sea water', *Nature*, **174**, 470–1.

Mason, B. J. (1971) *The Physics of Clouds* (2nd edn), Clarendon Press (Oxford), 671 pp.

Mason, B. J. (1973) 'The GARP Atlantic Tropical Experiment', *WMO Bull.*, **22**, 79–85.

Mason, B. J. (1975) *Clouds, Rain and Rainmaking* (2nd edn), Cambridge University Press, 189 pp.

Matheson, K. M. (1967) 'The meteorological effect of ice in the Gulf of St. Lawrence', *Arctic Met. Res. Group, McGill Univ., Publ. Met.*, No. 89, 110 pp.

Maury, M. F. (1844) 'Remarks on the Gulf Stream and currents of the sea', *Amer. J. Sci. Arts.*, **47**, 161–81.

Maury, M. F. (1851) *Explanations and sailing directions to accompany the wind and current charts,* C. Alexander (Washington), 351 pp.

Maury, M. F. (1855) *The Physical Geography of the Sea,* Harper (New York), 274 pp.

Maury, M. F. (1858) *The Physical Geography of the Sea and its Meteorology* (8th edn), Sampson Low (London), 485 pp.

Maury, M. F. (1963) *The Physical Geography of the Sea and its Meteorology* (Ed. J. Leighly), Harvard University (Belknap) Press, 432 pp.

McIntosh, D. H. (1972) *Meteorological Glossary*, Met. Off. (London), Her Majesty's Stationery Office (Met.O.842), 319 pp.

McIntosh, D. H. and Thom, A. S. (1969) *Essentials of Meteorology*, Wykeham Publications (London), 238 pp.

McPhee, M. G. and Smith, J. D. (1975) 'Measurements of the turbulent boundary under pack ice', *AIDJEX Bull.*, No. 29, University of Washington, Seattle, 49–92.

Medwin, H. (1970) 'In situ acoustic measurements of bubble populations in coastal ocean waters', *J. Geophys. Res.*, **75** (3), 599–611.

Meinardus, W. (1934) 'Eine neue Niederschlagskarte der Erde', *Petermanns Mitt.*, **80**, 1–4.

Meteorological Office (1976) 'Voluntary observing fleet and Ocean Weather Ships: Report of (Marine Division) work for 1975', *Mar. Obs.*, **46**, 50–4.

Michell, J. H. (1893) 'The highest waves in water', *Phil. Mag.* (Ser. 5), **36** (222), 430–7.

Miles, J. W. (1957) 'On the generation of surface waves by shear flows', *J. Fluid Mech.*, **3**, 185–204.

Mitchell, J. M. (1966) 'Stochastic models of air–sea interaction and climatic fluctuation', *Proc. Symp. Arctic Heat Budget and Atmos. Circ.* (Ed. J. O. Fletcher), The Rand Corporation, Santa Monica, Calif., 47–74.

Mohn, H. (1885) 'Die Strömungen des europoischen Nordmeeres', *Petermanns Mitt.* (Ergänzungsh.), **79**, 20 pp.

Monahan, E. C. (1971) 'Oceanic whitecaps', *J. Phys. Oceanogr.*, **1**, 139–44.

Montgomery, R. B. (1940) 'Observations of vertical humidity distribution above the ocean surface and their relation to evaporation', *Papers in phys. oceanogr. & met.* (Mass. Inst. Tech. and Woods Hole Oceanogr. Inst.), **7** (4), 30 pp.

Morgan, G. W. (1956) 'On the wind-driven ocean circulation', *Tellus*, **8**, 301–20.

Munk, W. H. (1950) 'On the wind-driven ocean circulation', *J. Met.*, **7**, 79–93.

Munk, W. H. (1957) Comments on review by C. L. Bretschneider, US Navy Dept. H.O. Publ. No. 603, *Trans. Amer. Geophys. Union*, **5** (38), 118–19.

Munk, W. H. and Carrier, G. F. (1950) 'The wind-driven circulation in ocean basins of various shapes', *Tellus*, **2**, 158–67.

Munk, W. H., Miller, G. R., Snodgrass, F. E. and Barber, N. F. (1963) 'Directional recording of swell from distant storms', *Phil. Trans. R. Soc.* (A), **255** (1062), 505–84.

Murray, R. and Ratcliffe, R. A. S. (1969) 'The summer weather of 1968: related atmospheric circulation and sea-temperature patterns', *Met. Mag.*, **98**, 201–19.

Namias, J. (1959) 'Recent seasonal interactions between North Pacific waters and the overlying atmospheric circulation', *J. Geophys. Res.*, **64** (6), 631–46.

Namias, J. (1964) 'Seasonal persistence of European blocking during 1958–60', *Tellus*, **16**, 394–407.

Namias, J. (1968) 'Long-range weather forecasting – history, current status and outlook', *Bull. Amer. Met. Soc.*, **49**, 438–70.

148

Namias, J. (1969) 'Seasonal interactions between the North Pacific Ocean and the atmosphere during the 1960's', *Mon. Wea. Rev., 97*, 173–92.

Namias, J. (1972) 'Large-scale and long-term fluctuations in some atmospheric and oceanic variables', *The Changing Chemistry of the Oceans (Nobel Symp.), 20*, 27–48.

Namias, J. (1973) 'Hurricane Agnes – an event shaped by large-scale air–sea systems generated during antecedent months', *Q. J. R. Met. Soc., 99*, 506–19.

Namias, J. (1974) 'Longevity of a coupled air–sea–continent system', *Mon. Wea. Rev., 102*, 638–48.

Namias, J. (1975) 'Stabilization of atmospheric circulation patterns by sea-surface temperature', *J. Mar. Res., 33* (Suppl.), 53–60.

Nansen, F. (1902) 'Oceanography of the North Polar Basin', *Norwegian North Polar Expedition, 1893–1896, Sci. Results, 3* (9), 427 pp.

National Academy of Sciences (1974) *U.S. Contribution to the Polar Experiment (POLEX): Part II POLEX– GARP (South),* National Academy of Sciences, Washington, DC, 33 pp.

National Academy of Sciences (1975) 'Numerical models of ocean circulation', *Proc. 1972 Durham (New Hampshire) Symp.* (Organized by Ocean Sci. Committee of Ocean Affairs Board), National Academy of Sciences, Washington, DC, 364 pp.

Neumann, G. (1968) *Ocean Currents,* Elsevier, 352 pp.

Neumann, G. and Pierson, W. J. (1957) 'A detailed comparison of theoretical wave spectra and wave forecasting methods', *Deut. Hydrogr. Zeit., 10*, 73–92, 134–46.

Neumann, G. and Pierson, W. J. (1963) 'Known and unknown properties of the frequency spectrum of a wind-generated sea', in *Ocean Wave Spectra,* Prentice-Hall, 9–21.

Neumann, G. and Pierson, W. J. (1966) *Principles of Physical Oceanography,* Prentice-Hall, 545 pp.

Newell, R. E., Vincent, D. G., Dopplick, T. G., Ferruzza, D. and Kidson, J. W. (1970) 'The energy balance of the global atmosphere', in *The Global Circulation of the Atmosphere* (Ed. G. A. Corby), Royal Meteorological Society (London), 42–90.

Newell, R. E. (1974) 'Changes in the poleward energy flux by the atmosphere and ocean as a possible cause for Ice Ages', *Quaternary Res., 4*, 117–27.

Newton, C. W., ed. (1972) *Meteorology of the Southern Hemisphere,* Amer. Met. Soc. Met. Monogr., 13 (35), 263 pp.

Nitta, T. (1975) 'GARP – observing systems simulation', *WMO Bull., 24*, 156–61.

Normand, Sir Charles (1953) 'Monsoon seasonal forecasting', *Q. J. R. Met. Soc., 79*, 463–73.

Nye, J. F. (1972) 'Glaciers – a physicist's view', *Proc. R. Inst. Gt. Britain, 45*, 255–81.

Nye, J. F. and Thomas, D. R. (1974) 'The use of satellite photographs to give the movement and deformation of sea ice', *AIDJEX Bull.,* No. 27, University of Washington, Seattle, 1–21.

Oerlemans, J. (1975) 'On the occurrence of Grosswetterlagen in winter related to anomalies in North Atlantic sea temperatures', *Met. Rund., 28*, 83–8.

Okuda, S. and Hayami, S. (1959) 'Experiments on evaporation from wavy water surface', *Rec. Oceanogr. Works in Japan, 5* (1), 6–13.

Ooyama, K. (1969) 'Numerical simulation of the life cycle of tropical cyclones', *J. Atmos. Sci., 26*, 3–40.

Östlund, H. G. (1968) 'Hurricane tritium II: air–sea exchange of water in Betsy 1965', *Tellus, 20*, 577–94.

Otto, L. (1973) 'Environmental factors in operations to combat oil spills', *WMO Mar. Sci. Affairs Rep. No. 9,* World Meteorological Organization (WMO – No. 359), 25 pp.

Palmén, E. (1948) 'On the formation and structure of tropical hurricanes', *Geophysica, 3*, 26–38.

Palmén, E. and Newton, C. W. (1969) *Atmospheric Circulation Systems,* Academic Press, 603 pp.

Parkhurst, P. G. (1955) 'Ocean meteorology: a century of scientific progress', *Mar. Obs., 25*, 16–21, 83–7.

Pasquill, F. (1972) 'Some aspects of boundary-layer description', *Q. J. R. Met. Soc., 98*, 469–94.

Pedgley, D. E. (1962) *A course in elementary meteorology,* Met. Off. (London), Her Majesty's Stationery Office (Met.O.707), 189 pp.

Pedlosky, J. (1975) 'The development of thermal anomalies in a coupled ocean–atmosphere model', *J. Atmos. Sci., 32*, 1501–14.

Perlroth, I. (1969) 'Effects of oceanographic media on equatorial Atlantic hurricanes', *Tellus, 21*, 230–44.

Perry, A. H. (1968) 'Turbulent heat flux patterns over the North Atlantic during recent winter months', *Met. Mag., 97*, 246–54.

Perry, A. H. (1972) 'June 1972 – the coldest June of the century', *Weather, 27*, 418–22.

Perry, A. H. (1975) 'Eastern North Atlantic sea-surface temperature anomalies and concurrent temperature and weather patterns over the British Isles', *Weather, 30*, 258–61.

Perry, J. D. (1968) 'Sea temperatures at OWS I', *Met. Mag., 97*, 33–43.

Petterssen, S., Bradbury, D. L. and Pedersen, K. (1962) 'The Norwegian cyclone models in relation to heat and cold sources', *Geofys. Publ.* (V. Bjerknes Centenary Volume), 24, 243–80.

Pettersson, O. (1904, 1907) 'On the influence of ice-melting upon oceanic circulation', *Geog. J., 24*, 285–333 and 30, 273–303.

Phillips, N. A. (1956) 'The general circulation of the atmosphere: a numerical experiment', *Q. J. R. Met. Soc., 82*, 123–64.

Phillips, N. A. (1966) 'Large-scale eddy motion in the western Atlantic', *J. Geophys. Res., 71* (16), 3883–91.

Phillips, O. M. (1957) 'On the generation of waves by turbulent wind', *J. Fluid Mech., 2*, 417–45.

Phillips, O. M. (1963) Comments on a paper by Darbyshire (1963), q.v., 33–8.

Phillips, O. M. (1969) *The Dynamics of the Upper Ocean,* Cambridge University Press, 261 pp.

Pierson, W. J. and Moskowitz, L. (1964) 'A proposed spectral form for fully-developed wind seas based on the similarity theory of S. A. Kitaigorodskii', *J. Geophys. Res., 69* (24), 5181–90.

Pierson, W. J., Neumann, G. and James, R. W. (1955) *Practical methods for observing and forecasting ocean waves by means of wave spectra and statistics,* US Navy Dept. H.O. Publ. No. 603 (Washington, DC).

Pillsbury, J. E. (1891) 'The Gulf Stream. A description of the methods employed in the investigation, and the results of the research', *Annual Rep. of the U.S. Coast & Geodetic Survey for 1890 (Appendix 10),* Washington, DC.

Pisharoty, P. R. (1965) 'Evaporation from the Arabian Sea and the Indian SW monsoon', *Proc. WMO/ UNESCO Symp. Met. Results IIOE* (Bombay), 43–54.

Postma, H. (1964) 'The exchange of oxygen and carbon dioxide between the ocean and the atmosphere', *Neth. J. Sea Res.*, **2**, 258–83.

Pounder, E. R. (1962) 'The physics of sea-ice', in *The Sea*, Vol. 1 (Ed. M. N. Hill), Interscience Publ. (Wiley), 826–38.

Prandle, D. (1975) 'Storm surges in the southern North Sea and River Thames', *Proc. R. Soc.* (A), **344** (1639), 509–39.

Prandtl, L. (1925) 'Bericht über Untersuchungen zur ausgebildeten Turbulenz', *Z. Angew, Math. Mech.*, **5**, 136.

Prescott, J. A. and Collins, J. A. (1951) 'The lag of temperature behind solar radiation', *Q. J. R. Met. Soc.*, **77**, 121–6.

Prestwich, J. (1875) 'Tables of temperatures of the sea at different depths beneath the surface, reduced and collated from the various observations made between the years 1749 and 1868, discussed', *Phil. Trans. R. Soc.*, **165**, 587–674.

Priestley, C. H. B. (1959) *Turbulent transfer in the lower atmosphere*, University of Chicago Press, 130 pp.

Priestley, C. H. B. (1964) 'Rainfall — sea-surface temperature associations on the New South Wales coast', *Australian Met. Mag.*, No. 47, 15–25.

Privett, D. W. (1960) 'The exchange of energy between the atmosphere and the oceans of the Southern Hemisphere', Met. Off. (London), *Geophys. Mem.*, No. 104 (M.O.631d), **13** (4), 61 pp.

Pyke, C. B. (1965) 'On the role of air–sea interaction in the development of cyclones', *Bull. Amer. Met. Soc.*, **46**, 4–15.

Quayle, R. G. (1974) 'A climatic comparison of ocean weather stations and transient ship records', *Mar. Wea. Log*, **18**, 307–11.

Quinn, W. H. and Burt, W. V. (1972) 'Use of the southern oscillation in weather prediction', *J. Appl. Met.*, **11**, 616–28.

Radok, U., Streten, N. and Weller, G. E. (1975) 'Atmosphere and ice', *Oceanus*, **7**, 17–28.

Ramage, C. S. (1965) *Meteorology in the Indian Ocean*, World Meteorological Organization (WMO – No. 166), 31 pp.

Ramage, C. S. (1966) 'The summer atmospheric circulation over the Arabian Sea', *J. Atmos. Sci.*, **23**, 144–50.

Ramage, C. S. (1971) *Monsoon Meteorology*, Academic Press, 296 pp.

Ramage, C. S. (1972) 'Interaction between tropical cyclones and the China Seas', *Weather*, **27**, 484–94.

Ramage, C. S. (1974) 'Monsoonal influences on the annual variation of tropical cyclone development over the Indian and Pacific Oceans', *Mon. Wea. Rev.*, **102**, 745–53.

Ramage, C. S. (1975) 'Preliminary discussion of the meteorology of the 1972–73 El Niño', *Bull. Amer. Met. Soc.*, **56**, 234–42.

Rao, P. K., Smith, W. L. and Koffler, R. (1972) 'Global sea-surface temperature distribution determined from an environmental satellite', *Mon. Wea. Rev.*, **100**, 101–14.

Rasool, S. I. and Prabhakara, C. (1965) *Radiation studies from meteorological satellites*, NY Univ., Dept. Met. Oceanogr., Geophys. Sci. Lab. Rep. 65–1.

Rasool, S. I. and Prabhakara, C. (1966) 'Heat budget of the Southern Hemisphere', in *Problems of atmospheric circulation* (Eds. R. V. Garcia and T. F. Malone), Spartan Books (New York), 76–92.

Ratcliffe, R. A. S. (1971) 'North Atlantic sea temperature classification, 1877–1970', *Met. Mag.*, **100**, 225–32.

Ratcliffe, R. A. S. (1973) 'Recent work on sea-surface temperature anomalies related to long-range forecasting', *Weather*, **28**, 106–17.

Ratcliffe, R. A. S. and Murray, R. (1970) 'New lag associations between North Atlantic sea temperature and European pressure applied to long-range weather forecasting', *Q. J. R. Met. Soc.*, **96**, 226–46.

Redfield, W. C. (1834) 'Summary statements of some of the leading facts in meteorology', *Amer. J. Sci. Arts*, **25**, 122–35.

Ricker, W. E. (1969) 'Food from the sea', in *Resources and Man* (Ed. P. Cloud *et al.*), Freeman & Co., 87–108.

Riehl, H. (1969) 'On the rôle of the tropics in the general circulation of the atmosphere', *Weather*, **24**, 288–308.

Riehl, H. and Malkus, J. S. (1957) 'On the heat balance and maintenance of circulation in the trades', *Q. J. R. Met. Soc.*, **83**, 21–9.

Riehl, H. and Malkus, J. S. (1958) 'On the heat balance in the equatorial trough zone', *Geophysica*, **6**, 503–38.

Riehl, H., Yeh, T. C., Malkus, J. S. and LaSeur, N. E. (1951) 'The north-east trade of the Pacific Ocean', *Q. J. R. Met. Soc.*, **77**, 598–626.

Roberts, E. D. (1971) *Handbook of Aviation Meteorology*, Met. Off. (London), Her Majesty's Stationery Office (Met.O.818), 404 pp.

Robin, G. de Q. (1966) 'Pack ice and waves', *Proc. Symp. Antarctic Oceanogr.* (Santiago), 191–7.

Robinson, A. R. and Stommel, H. (1959) 'The oceanic thermocline and associated thermohaline circulation', *Tellus*, **11**, 295–308.

Robinson, A. R., Luyten, J. R. and Fuglister, F. C. (1974) 'Transient Gulf Stream meandering. Part I: an observational experiment', *J. Phys. Oceanogr.*, **4**, 237–55.

Robinson, A. R., Tomasin, A. and Artegiani, A. (1973) 'Flooding of Venice: phenomenology and prediction of the Adriatic storm surge', *Q. J. R. Met. Soc.*, **99**, 688–92.

Robinson, G. D. (1966) 'Another look at some problems of the air–sea interface', *Q. J. R. Met. Soc.*, **92**, 451–65.

Rodewald, M. (1972) 'Some hydroclimatic characteristics of the decade 1961–70 in the North Atlantic and in the Arctic', *Deut. Hydrogr. Zeit.*, **25**, 97–117.

Rodhe, B. (1962) 'The effect of turbulence on fog formation', *Tellus*, **14**, 49–86.

Roll, H. U. (1965) *Physics of the Marine Atmosphere*, Academic Press, 426 pp.

Roll, H. U. (1972) 'Problem areas in air–sea interaction', in *Studies in Physical Oceanography* (Wüst 80th birthday tribute, Ed. A. L. Gordon), publ. by Gordon & Breach, Vol. 1, 63–72.

Rosenthal, S. C. (1960) 'The interdiurnal variability of surface air temperature over the North Atlantic Ocean', *J. Met.*, **17**, 1–7, 78–83.

Rossby, C.-G. (1936) 'Dynamics of steady ocean currents in the light of experimental fluid mechanics', *Papers in phys. oceanogr. & met.* (Mass. Inst. Tech. and Woods Hole Oceanogr. Inst.), **5** (1), 43 pp.

Rossby, C.-G. (1956) 'Current problems in meteorology', in *The Atmosphere and the Sea in Motion* (Ed. B. Bolin), Rockefeller Inst. Press (1959), 9–50; trans. of 'Aktuella meteorologiska problem', *Svensk Naturvetenskap* (1956), 15–80.

Rossiter, J. R. (1954) 'The North Sea storm surge of 31 January and 1 February 1953', *Phil. Trans. R. Soc.* (A), **246** (915), 371–400.

Rowntree, P. R. (1972) 'The influence of tropical east Pacific Ocean temperatures on the atmosphere', *Q. J. R. Met. Soc.,* **98**, 290–321.

Rowntree, P. R. (1976a) 'Response of the atmosphere to a tropical Atlantic ocean temperature anomaly', *Q. J. R. Met. Soc.,* **102**, 607–25.

Rowntree, P. R. (1976b) 'Tropical forcing of atmospheric motions in a numerical model', *Q. J. R. Met. Soc.,* **102**, 583–605.

Royal Society (1975) *Air–Sea Interaction Project* (Plans for the Joint Air–Sea Interaction Experiments JASIN 1977 and 1978), The Royal Society of London, 107 pp.

Rubin, M. J. (1958) 'An occurrence of steam fog in Antarctic waters', *Weather,* **13**, 235–8.

Ryder, C. (1917) *Monthly mean temperatures of the surface waters in the Atlantic north of 50°N,* Publ. Danske Met. Inst. Aarbager, Copenhagen.

Sabine, E. (1846) 'On the cause of remarkably mild winters which occasionally occur in England', *Phil. Mag.* (Third Series), **28**, 317–24.

Saha, K. (1970) 'Air and water vapour transport across the equator in western Indian Ocean during northern summer', *Tellus,* **22**, 681–7.

Saha, K. (1971) 'Mean cloud distributions over tropical oceans', *Tellus,* **23**, 183–94.

Saha, K. and **Bavadekar, S. N.** (1973) 'Water vapour budget and precipitation over the Arabian Sea during the northern summer', *Q. J. R. Met. Soc.,* **99**, 273–8.

Sanderson, R. M. (1975) 'Changes in the area of Arctic sea-ice 1966 to 1974', *Met. Mag.,* **104**, 313–23.

Sandström, J. W. (1908) 'Dynamische Versuche mit Meerwasser', *Ann. Hydrogr. Marit. Met.,* **36**, 6–23.

Saunders, P. M. (1964) 'Sea smoke and steam fog', *Q. J. R. Met. Soc.,* **90**, 156–65.

Saur, J. F. T. (1963) 'A study of the quality of sea-water temperature records in logs of ships' weather observations', *J. Appl. Met.,* **2**, 417–21.

Schell, I. I. (1971) 'On mathematical and "natural" models for the study of climate changes', *J. Appl. Met.,* **10**, 1344–6.

Schell, I. I. and **Corkum, D. A.** (1976) 'On a thermal lag in the North Atlantic Ocean during a climatic change', *J. Phys. Oceanogr.,* **6**, 125–9.

Schneider, S. H. and **Dickinson, R. E.** (1974) 'Climate modelling', *Rev. Geophys. Space Phys.,* **12**, 447–93.

Schneider, S. H. and **Washington, W. M.** (1973) 'Cloudiness as a global climatic feedback mechanism' (abstract only), *Bull. Amer. Met. Soc.,* **54**, 742.

Schroeder, E. H. (1963) *North Atlantic temperature at a depth of 200 metres,* Serial Atlas of the Marine Environment, Folio 2, Amer. Geog. Soc.

Schwerdtfeger, W. (1975) 'The effect of the Antarctic Peninsula on the temperature regime of the Weddell Sea', *Mon. Wea. Rev.,* **103**, 45–51.

Scorer, R. S. (1958) *Natural Aerodynamics,* Pergamon, 312 pp.

Scorer, R. S. and **Ludlam, F. H.** (1953) 'Bubble theory of penetrative convection', *Q. J. R. Met. Soc.,* **79**, 94–103.

Scripps (1975) 'NORPAX', in *Ann. Rep. for year ending June 30, 1974* (Scripps Inst. Oceanogr., San Diego), **8** (3), 34–5.

Selby, M. J. (1973) 'Antarctica: the key to the Ice Age', *New Zealand Geographer,* **29**, 134–50.

Sellers, W. D. (1965) *Physical Climatology,* University of Chicago Press, 272 pp.

Sethuraman, S. and **Raynor, G. S.** (1975) 'Surface drag coefficient dependence on the aerodynamic roughness of the sea', *J. Geophys. Res.,* **80** (36), 4983–8.

Shaw, Sir Napier (1923) *Forecasting Weather,* Constable, 584 pp.

Shaw, Sir Napier (1926) *Manual of Meteorology* (Vol. 1, Meteorology in History), Cambridge University Press, 339 pp.

Shellard, H. C. (1975) 'Lerwick anemograph records 1957–70 and the offshore industry', *Met. Mag.,* **104**, 189–208.

Sheppard, P. A. (1968) 'Global atmospheric research', *Weather,* **23**, 262–83.

Sheppard, P. A. (1970) 'The atmospheric boundary layer in relation to large-scale dynamics', in *The Global Circulation of the Atmosphere* (Ed. G. A. Corby), Royal Meteorological Society (London), 91–112.

Shukla, J. (1975) 'Effect of Arabian Sea-surface temperature anomaly on Indian summer monsoon: a numerical experiment with the GFDL model', *J. Atmos. Sci.,* **32**, 503–11.

Simpson, R. H., Frank, N., Shideler, D. and **Johnson, H. M.** (1968) 'Atlantic tropical disturbances, 1967', *Mon. Wea. Rev.,* **96**, 251–9.

Simpson, R. H., Frank, N., Shideler, D. and **Johnson, H. M.** (1969) 'Atlantic tropical disturbances of 1968', *Mon. Wea. Rev.,* **97**, 240–55.

Smagorinsky, J. (1970) 'Numerical simulation of the global atmosphere', in *The Global Circulation of the Atmosphere* (Ed. G. A. Corby), Royal Meteorological Society (London), 24–41.

Smagorinsky, J., Manabe, S. and **Holloway, J. L.** (1965) 'Numerical results from a nine-level general circulation model of the atmosphere', *Mon. Wea. Rev.,* **93**, 727–68.

Smed, J. (1947–60 and 1962–65) 'Monthly anomalies of the surface temperature in the sea', *Ann. Biol.* (Copenhagen), **1** to **17**.

Smith, C. L., Zipser, E. J., Daggupaty, S. M. and **Sapp, L.** (1975) 'An experiment in tropical mesoscale analysis', *Mon. Wea. Rev.,* **103**, 878–92, 893–903.

Smith, F. B. (1975) 'Turbulence in the atmospheric boundary layer', *Sci. Prog.,* **62**, 127–51.

Smith, R. L. (1968) 'Upwelling', in *Oceanography: contemporary readings in ocean sciences* (Ed. R. G. Pirie), Oxford University Press (1973), 126–47.

Smith, S. D. and **Banke, E. G.** (1975) 'Variation of the sea-surface drag coefficient with wind speed', *Q. J. R. Met. Soc.,* **101**, 665–73.

Smith, W. L. and others (1970) 'The determination of sea-surface temperature anomalies from satellite high resolution infra-red window radiation measurements', *Mon. Wea. Rev.,* **98**, 604–11.

Snodgrass, F. E. and 5 co-authors (1966) 'Propagation of ocean swell across the Pacific', *Phil. Trans. R. Soc.* (A), **259** (1103), 431–97.

Snyder, R. L. (1974) 'A field study of wave-induced

151

pressure-fluctuations above surface gravity waves',
J. Mar. Res., 32, 497–531.

Spar, J. (1965) 'Air–sea exchange as a factor in synoptic-
scale meteorology in middle latitudes', *U.S. Dept.
Commerce/ESSA Tech. Note 9 – SAIL – 1*, 1–16.

Spar, J. (1973) 'Some effects of surface anomalies in a
global generated circulation model', *Mon. Wea. Rev.,
101*, 91–100; 'Transequatorial effects of sea-surface
temperature anomalies in a global generated model',
Mon. Wea. Rev., 101, 554–63, 767–73.

Spar, J. and **Atlas, R.** (1975) 'Atmospheric response to
variations in sea-surface temperature', *J. Appl. Met.,
14*, 1235–45.

Stanton, Sir T. E., Marshall, D. and **Houghton, R.** (1932)
'The growth of waves on water due to the action of
the wind', *Proc. R. Soc. (A), 137*, 283–93.

Starbuck, L. (1953) 'Arctic sea-smoke at Hong Kong',
Weather, 8, 77–8.

Steers, J. A. (1953) 'The East Coast floods of January 31–
February 1 1953', *Geog. J., 119*, 280–98.

Stevenson, C. M. (1968) 'The snowfalls of early December
1967', *Weather, 23*, 156–61.

Stevenson, R. E. (1964) 'The influence of a ship on the
surrounding air and water temperatures', *J. Appl. Met.,
3*, 115–18.

Stewart, R. W. (1967) 'The atmosphere and the ocean', in
Oceanography: readings from Scientific American (Ed.
J. R. Moore), Freeman & Co. (1971), 35–44.

Stewart, R. W. (1974) 'The air–sea momentum exchange',
Boundary-Layer Met., 6, 151–67.

Stewart, R. W. (1975) 'Atmospheric boundary layer',
WMO Bull., 24, 82–6.

Stoddart, D. R. (1971) 'Rainfall on Indian coral islands',
Atoll Res. Bull., No. 147, Smithsonian Inst.
(Washington, DC), 21 pp.

Stokes, G. G. (1880) Supplement to a paper on the theory
of oscillatory waves, in *Mathematical and Physical
Papers*, Vol. 1, Cambridge University Press, 314–26.

Stommel, H. (1948) 'The westward intensification of
wind-driven ocean currents', *Trans. Amer. Geophys.
Union, 29*, 202–6.

Stommel, H. (1957) 'A survey of ocean current theory',
Deep-Sea Res., 4, 149–84.

Stommel, H. (1958) 'The abyssal circulation', *Deep-Sea
Res., 5*, 80–2.

Stommel, H. (1962) 'On the smallness of sinking regions
in the ocean', *Proc. Nat. Acad. Sci., 48*, 766–72.

Stommel, H. (1965) *The Gulf Stream: a physical and
dynamical description* (2nd edn), University of
California and Cambridge University Press, 248 pp.

Stommel, H. and **Arons, A. B.** (1960) 'On the abyssal
circulation of the world ocean', *Deep-Sea Res., 7*,
140–54, 217–33.

Stommel, H. and **Yoshida, K.**, eds. (1972) *Kuroshio:
Physical Aspects of the Japan Current*, University of
Washington Press, 517 pp.

Strickland, J. D. H. (1958) 'Solar radiation penetrating
the ocean. A review of requirements, data and
methods of measurement, with particular reference to
photosynthetic productivity', *J. Fisheries Res. Bd.
Canada, 15*, 453–93.

Suomi, V. E. (1970) 'Recent developments in satellite
techniques for observing and sensing the atmosphere',
in *The Global Circulation of the Atmosphere* (Ed.
G. A. Corby), Royal Meteorological Society (London),
222–34.

Supan, A. (1898) 'Die Verteilung der jährlichen
Niederschlagshöhe im Atlantischen und Indischen
Ozean', *Petermanns Mitt., 44*, 179–82, Taf. 13.

Sverdrup, H. U. (1937) 'On the evaporation from the
oceans', *J. Mar. Res., 1*, 3–14.

Sverdrup, H. U. (1943) 'On the ratio between heat
conduction from the sea surface and heat used for
evaporation', *Ann. N.Y. Acad. Sci., 44*, 81–8.

Sverdrup, H. U. (1945) *Oceanography for Meteorologists*,
Allen & Unwin, 246 pp.

Sverdrup, H. U. (1947) 'Wind-driven currents in a
baroclinic ocean; with application to the equatorial
currents of the eastern Pacific', *Proc. Nat. Acad. Sci.,
33*, 318–26.

Sverdrup, H. U. (1951) 'Evaporation from the oceans', in
Compendium of Meteorology (Ed. T. F. Malone),
Amer. Met. Soc., 1071–81.

Sverdrup, H. U. (1957) 'Oceanography', in *Handbuch der
Physik, 48*, Springer-Verlag, 608–70.

Sverdrup, H. U. and **Munk, W. H.** (1947) *Wind, Sea, and
Swell: Theory of Relations for Forecasting*, US Navy
Dept. H.O. Publ. No. 601, 44 pp.

Swallow, J. C. and **Bruce, J. G.** (1966) 'Current
measurements off the Somali coast during the
southwest monsoon of 1964', *Deep-Sea Res., 13*,
861–88.

Swinbank, W. C. (1951) 'The measurement of vertical
transfer of heat and water vapour and momentum in
the lower atmosphere with some results', *J. Met., 8*,
135–45.

Swinbank, W. C. (1955) 'An experimental study of eddy
transports in the lower atmosphere', *C.S.I.R.O.
(Australia) Div. Met. Phys. Tech. Pap.*, No. 2.

Swinbank, W. C. (1968) 'A comparison between
predictions of dimensional analysis for the constant
flux layer and observations in unstable conditions',
Q. J. R. Met. Soc., 94, 460–7.

Swithinbank, C. and **Zumberge, J. H.** (1965) 'The ice
shelves', in *Antarctica* (Ed. T. Hatherton), Methuen,
199–220.

Tessan, U. de (1844) *Voyage Autour du Monde sur la
Frégate "La Vénus", pendant les années 1836–1839*,
Paris, 10 vols. (N.B. The British Library has been
unable to trace in the United Kingdom a loan copy of
the accompanying *Carte des Courants et des
Températures de l'Eau à la surface de la Mer, observés
à bord de "La Vénus" en 1837, 38 et 39*.)

Thompson, B. (1798) 'On the propagation of heat in
fluids', in *Essays, political, economical, and
philosophical* (publ. London, 1800), Vol. 2, 197–386.

Thomson, C. W. (1871) 'On the distribution of
temperature in the North Atlantic', *Nature, 4*, 251–3.

Thoulet, J. (1928) 'Le courant de Humboldt et la mer de
l'Île de Pâques', *Ann. Inst. Océanogr., 5* (fasc. II),
1–12.

Townsend, J. (1975) 'Forecasting "negative" storm surges
in the southern North Sea', *Mar. Obs., 45*, 27–35.

Tsunogai, S. (1975) 'Sea salt particles transported to the
land', *Tellus, 27*, 51–8.

Turner, J. S. and **Kraus, E. B.** (1967) 'A one-dimensional
model of the seasonal thermocline', *Tellus, 19*,
88–105.

Ufford, H. A. Q. v. (1953) 'The disastrous storm surge of
1 February', *Weather, 8*, 116–20.

Untersteiner, N. (1963) 'Ice budget of the Arctic Ocean',
Proc. Arctic Basin Symp. (Arctic Inst. N. Amer.,

Washington), 219–30.

Ursell, F. (1956) 'Wave generation by wind', in *Surveys in Mechanics* (Eds. G. K. Batchelor and R. M. Davies), Cambridge University Press, 216–49.

US Navy (1966) *Atlas of Bathythermograph Data* – Indian Ocean, Nat. Oceanogr. Data Centre, Washington.

Vacnadze, D. I., Dujceva, M. A. and **Ped, D. A.** (1970) 'Interrelation between water temperature and air temperature fields in the North Atlantic', *Leningrad Gidromet. Nauc. Issled. Cent. SSSR,* T Vyp 63, 43–56.

Veronis, G. and **Stommel, H.** (1956) 'The action of variable wind stresses on a stratified ocean', *J. Mar. Res.,* **15**, 43–75.

Vines, R. G. (1959) 'Wind stress on a water surface: measurements at low wind speeds with the aid of surface films', *Q. J. R. Met. Soc.,* **85**, 159–62.

Vladimirov, O. A. and **Nikolaev, J. V.** (1970) 'Some physico-statistical parameters of the water and air temperature fields in the North Atlantic', *AIDJEX Bull.,* No. 10, University of Washington, Seattle, 165–76.

Vonder Haar, T. H. and **Suomi, V. E.** (1969) 'Satellite observations of the earth's radiation budget', *Science,* **163** (3868), 667–9.

Vowinckel, E. (1965) 'The energy budget over the North Atlantic Ocean – January 1963', *Arctic Met. Res. Group, McGill Univ., Publ. Met.,* No. 78.

Vowinckel, E. and **Orvig, S.** (1966) 'Energy balance of the Arctic. V The heat budget over the Arctic Ocean', *Archiv für Met., Geophys. und Bioklim. (B),* **14**, 303–25.

Vowinckel, E. and **Orvig, S.** (1970) 'The climate of the North Polar Basin', in *World Survey of Climatology,* Vol. 14 (Ed. S. Orvig), Elsevier, 129–252;

Wahl, E. W. and **Bryson, R. A.** (1975) 'Recent changes in Atlantic surface temperatures', *Nature,* **254**, 45–6.

Walden, H. (1963) 'Comparison of one-dimensional wave spectra recorded in the German Bight with various "theoretical" spectra', in *Ocean Wave Spectra,* Prentice-Hall, 67–81.

Walker, J. M. (1972a) 'Monsoons and the global circulation', *Met. Mag.,* **101**, 349–55.

Walker, J. M. (1972b) 'Convective processes and the summer monsoon of southern Asia', *Vayu Mandal (Bull. Indian Met. Soc.),* **2**, 169–73.

Walker, J. M. (1975) 'On summer atmospheric processes over south-west Asia', *Tellus,* **27**, 491–6.

Walker, J. M. and **Penney, P. W.** (1973) 'Arctic sea-ice and maritime transport technology', *Weather,* **28**, 358–71.

Warren, B. A. (1963) 'Topographic influences on the path of the Gulf Stream', *Tellus,* **15**, 167–83.

Warren, B. A. (1966) 'Medieval Arab references to the seasonally reversing currents of the North Indian Ocean', *Deep-Sea Res.,* **13**, 167–71.

Warren, B. A., Stommel, H. and **Swallow, J. C.** (1966) 'Water masses and patterns of flow in the Somali Basin during the southwest monsoon of 1964', *Deep-Sea Res.,* **13**, 825–60.

Weeks, W. F. (1966) 'Understanding the variations of the physical properties of sea ice', *Proc. Symp. Antarctic Oceanogr.* (Santiago), 173–90.

Weeks, W. F. and **Assur, A.** (1967) *The mechanical properties of sea ice,* US Army Cold Regions Res. & Eng. Lab. monogr., DA Project 1VO25001A130, 80 pp.

Weeks, W. F., Kovacs, A. and **Hibler, W. D.** (1971)

'Pressure ridge characteristics in the Arctic coastal environment', *Proc. Conf. Port & Ocean Eng. under Arctic Conditions* (Norway), Vol. 1, 152–83.

Weller, G. and **Bierly, E. W.** (1973) 'The Polar Experiment (POLEX)', *Bull. Amer. Met. Soc.,* **54**, 212–18.

Weyl, P. K. (1968) 'The rôle of the oceans in climatic change: a theory of the ice ages', *Met. Monogr.,* **8** (30), 37–62.

White, W. B. and **Barnett, T. P.** (1972) 'A servomechanism in the ocean/atmosphere system of the mid-latitude North Pacific', *J. Phys. Oceanogr.,* **2**, 372–81.

White, W. B. and **Clarke, N. E.** (1975) 'On the development of blocking ridge activity over the central North Pacific', *J. Atmos. Sci.,* **32**, 489–502.

Wiener, N. (1968) 'Cybernetics in history', in *Modern Systems Research for the Behavioural Scientist* (Ed. W. Buckley), Aldine Publ. Co. (Chicago), 31–6.

Williams, J. (1799) *Thermometrical Navigation,* Philadelphia.

Williams, J., Higginson, J. J. and **Rohrbough, J. D.** (1973) *Sea & Air: The Marine Environment* (2nd edn), Naval Institute Press (Annapolis), 338 pp.

Wilson, B. W. (1960) 'Note on surface wind stress over water at low and high wind speeds', *J. Geophys. Res.,* **65** (10), 3377–82.

Winston, J. S. (1955) 'Physical aspects of rapid cyclogenesis in the Gulf of Alaska', *Tellus,* **7**, 481–500.

Winston, J. S. (1969) 'Global distribution of cloudiness and radiation as measured from weather satellites', in *World Survey of Climatology,* Vol. 4 (Ed. D. F. Rex), Elsevier, 247–80.

Witte, E. (1880) 'Das Emporquellen von kaltem Wasser an meridionalen Küsten', *Ann. Hydrogr. Berlin,* **8**, 192–3.

Wittmann, W. I. and **Schule, J. J.** (1966) 'Comments on the mass budget of Arctic pack-ice', *Proc. Symp. Arctic Heat Budget and Atmos. Circ.* (Ed. J. O. Fletcher), The Rand Corporation, Santa Monica, Calif., 215–46.

WMO (1954) *Methods of Observation at Sea,* World Meteorological Organization Tech. Note No. 2 (WMO – No. 26), 35 pp.

WMO (1962a) *Precipitation Measurements at Sea,* World Meteorological Organization Tech. Note No. 47 (WMO – No. 124), 18 pp.

WMO (1962b) *Climatological normals (CLINO) for CLIMAT and CLIMAT SHIP stations for the period 1931–1960,* World Meteorological Organization (WMO – No. 117).

WMO (1966) 'The global data-processing system and meteorological service to aviation', *World Weather Watch Planning Report,* No. 13, World Meteorological Organization, 40 pp.

WMO (1969) 'Meteorological observations from mobile and fixed ships', *World Weather Watch Planning Report,* No. 7 (first publ. 1966), World Meteorological Organization, 24 pp.

WMO (1976) 'Global Atmospheric Research Programme', *WMO Bull.,* **25**, 32–8.

WMO/ICSU (1974) 'Report of the meeting on drifting buoys for the first GARP Global Experiment', *GARP Special Report,* No. 13, World Meteorological Organization and International Council of Scientific Unions, 80 pp.

WMO Secretariat (1966) *The Essential Elements of the World Weather Watch,* World Meteorological Organization, 25 pp.

WMO Secretariat (1967) *World Weather Watch: The Plan and Implementation Programme*, World Meteorological Organization, 56 pp.

WMO Secretariat (1970) 'Global ocean research', *WMO Mar. Sci. Affairs Rep.*, No. 1, World Meteorological Organization, 47 pp.

WMO Secretariat (1971) *World Weather Watch: The Plan and Implementation Programme*, World Meteorological Organization (WMO – No. 296), 82 pp.

WMO Secretariat (1973) *Joint IOC/WMO Planning Group for IGOSS (Integrated Global Ocean Station System). Report of Second Session, Geneva, 13–17 August 1973*, publ. by UNESCO, the Intergovernmental Oceanographic Commission and the World Meteorological Organization, 60 pp.

Wolff, P. M. and **Cartensen, L. P.** (1965) 'Analysis and forecasting of sea-surface temperature', *Fleet Num. Wea. Fac.* (Monterey, Calif.) *Tech. Note*, No. 8, 48 pp.

Woodcock, A. H. (1940) 'Convection and soaring over the open sea', *J. Mar. Res.*, **3**, 248–53.

Woodcock, A. H. (1975) 'Thermals over the sea and gull flight behaviour', *Boundary-Layer Met.*, **9**, 63–8.

Woodcock, A. H. and **Wyman, J.** (1947) 'Convective motion in air over the sea', *Annals N.Y. Acad. Sci.*, **48**, 749–76.

Woods, J. D. (1968a) 'CAT under water', *Weather*, **23**, 224–35.

Woods, J. D. (1968b) 'An investigation of some physical processes associated with the vertical flow of heat through the upper ocean', *Met. Mag.*, **97**, 65–72.

Woodwell, G. M. (1967) 'Toxic substances and ecological cycles', in *Man and the Ecosphere* (Ed. P. R. Ehrlich et al.), Freeman & Co., 128–35.

Wooster, W. S. and **Guillen, O.** (1974) 'Characteristics of El Niño 1972', *J. Mar. Res.*, **32**, 387–404.

Worthington, L. V. (1977) *On the North Atlantic Circulation*, Johns Hopkins Press (Oceanogr. Studs. No. 6), 112 pp.

Wu, J. (1969) 'Wind stress and surface roughness at air–sea interface', *J. Geophys. Res.*, **74** (2), 444–55.

Wu, J. (1971) 'Observations on long waves sweeping through short waves', *Tellus*, **23**, 364–9.

Wüst, G. (1936) 'Oberflächensalzgehalt, Verdunstung und Niederschlag auf dem Weltmeere', *Ländenkundl. Forschung*, Festschrift N. Krebs, 347–59.

Wüst, G. (1968) 'History of investigations of the longitudinal deep-sea circulation (1800–1922)', *Bull. Inst. Océanogr. Monaco*, No. Spécial 2, 109–20.

Wyrtki, K. (1961) 'The thermohaline circulation in relation to the general circulation in the oceans', *Deep-Sea Res.*, **8**, 39–64.

Wyrtki, K. (1962) 'The upwelling in the region between Java and Australia during the south-east monsoon', *Australian J. Mar. Freshwater Res.*, **13**, 217–25.

Wyrtki, K. (1963) 'The horizontal and vertical field of motion in the Peru Current', *Bull. Scripps Inst. Oceanogr.*, **8** (4), 313–46.

Wyrtki, K. (1965) 'The average annual heat balance of the North Pacific Ocean and its relation to ocean circulation', *J. Geophys. Res.*, **70** (18), 4547–59.

Yi-Yuan Yu (1952) 'Breaking of waves by an opposing current', *Trans. Amer. Geophys. Union*, **33**, 39–41.

Zöppritz, K. (1878) 'Hydrodynamische Probleme in Beziehung zur Theorie der Meeresströmungen', *Wied. Ann.*, **3**, 582.

Index